The Sustainability Conundrum

The Sustainability Conundrum: Balancing Energy Transition and Global Development investigates how a clean environment and mitigation of Global Climate Change will affect the UN's sustainability goals for development, poverty elimination, and inequality reduction. The book covers the necessity of energy use for humans, the global demand and the supply of energy, and the environmental challenges of energy consumption.

Outlining the interconnections between four of the UN goals, the book discusses the Global Climate Change problem by explaining how the scientific community eradicated other global environmental threats, such as acid rain, lead contamination, and the ozone hole. It establishes the uniqueness of Global Climate Change; evaluates proposed methods and "roadmaps" to achieve meaningful carbon dioxide emissions reductions; computes the cost of the necessary infrastructure for their implementation on a regional and a global scale; and draws conclusions about the implementation of the proposed energy transition and the sustainability goals for the elimination of poverty and reduction of inequalities within and among nations.

This book is intended for students of sustainability initiatives and any readers interested in the ongoing challenges surrounding sustainability and global economic development.

Efstathios E. Michaelides is the Tex Moncrief Chair of Engineering at Texas Christian University (TCU), USA. Among his previous positions, he was the Founder and Director of the NSF-supported Center on Simulation, Visualization and Real Time Computing (SiViRT) at the University of Texas, San Antonio, and the director of the South Central Center of the National Institute for Global Environmental Change (2002–2007) at Tulane University. Professor Michaelides holds a B.A. degree (honors) from Oxford University, UK, and M.S. and Ph.D. degrees from Brown University, USA. He has published more than 200 journal papers; has contributed more than 350 papers and presentations in national and international conferences; and has authored 9 books on energy, the environment, and multiphase flow. Among other honors, he has received the ASME Fluids Engineering Award (2014) and the ASME James Harry Potter Medal (2024) "for eminent contributions in the science of Thermodynamics." Professor Michaelides is currently a fellow of ASME and the ASTFE.

The Sustainability Conundrum

Balancing Energy Transition and Global Development

Efstathios E. Michaelides

CRC Press
Taylor & Francis Group
Boca Raton London New York

CRC Press is an imprint of the
Taylor & Francis Group, an **informa** business

Designed cover image: Shutterstock

First edition published 2027
by CRC Press
2385 NW Executive Center Drive, Suite 320, Boca Raton FL 33431

and by CRC Press
4 Park Square, Milton Park, Abingdon, Oxon, OX14 4RN

CRC Press is an imprint of Taylor & Francis Group, LLC

ISBN: 978-1-041-14483-0 (hbk)
ISBN: 978-1-041-14480-9 (pbk)
ISBN: 978-1-003-67465-8 (ebk)

DOI: 10.1201/9781003674658

Typeset in Times
by KnowledgeWorks Global Ltd.

Contents

About the Author

Stathis Michaelides is the Tex Moncrief Chair of Engineering at Texas Christian University (TCU). Prior to this, he was chair of the Department of Mechanical Engineering of the University of Texas at San Antonio, and was the Founder and Director of the NSF-supported Center on Simulation, Visualization and Real Time Computing (SiViRT). In the past he was the Founding Chair of the Department of Mechanical and Energy Engineering at the University of North Texas; the Leo S. Weil Professor of Mechanical Engineering at Tulane University; Director of the South-Central Center of the National Institute for Global Environmental Change (2002–2007) at Tulane University where he also served as the Associate Dean for Graduate Studies and Research and Head of the Mechanical Engineering Department. Between 1980 and 1989, he was on the faculty of the University of Delaware, where he also served as Acting Chairman of the Mechanical Engineering Department.

Michaelides holds a B.A. degree (honors) from Oxford University and M.S. and Ph.D. degrees from Brown University. He was awarded an honorary M.A. degree from Oxford University (1983); the Casberg and Schillizzi Scholarships at St. Johns College, Oxford; the student chapter ASME/Phi, Beta, Tau excellence in teaching award (1991 and 2001); the Lee H. Johnson award for teaching excellence (1995); a Senior Fulbright Fellowship (1997); the ASME Freeman Scholar Award (2002); the Outstanding Researcher Award at Tulane (2003); the ASME Outstanding Service Award (2007); the ASME Fluids Engineering Award (2014); the ASME 90th Anniversary of FED Medal, 2016; the ASME Edwin E. Church Medal (2021), "for the development of several noteworthy Mechanical Engineering programs, and for significant outreach efforts to increase diversity in Mechanical Engineering education"; and the ASME James Harry Potter Medal (2024) "for eminent contributions in the science of Thermodynamics."

Michaelides served as chair of the ASME Fluids Engineering Division in 2005-2006. He served twice as the President of the ASEE Gulf-South Region (1992–1993 and 2016–2017). He was the chair of the 4th International Conference on Multiphase Flows (New Orleans, 2001) and was the vice-chair of the 5th International Conference on Multiphase Flows (2004). He has published more than 200 journal papers and has contributed more than 350 papers and presentations in national and international conferences. He is currently editor of the *Journal of Non-Equilibrium Thermodynamics*. He has also authored nine other books on Energy, Sustainability, and Multiphase Flow.

List of Abbreviations

A/C	Air-Conditioning
AI	Artificial Intelligence
BRICS	Brazil, Russia, India, China, South Africa
CAES	Compressed Air Energy Storage
CBAM	Carbon Border Adjustment Mechanism
CCS	Carbon Capture and Sequestration
COP	Conference of Parties
EPA	Environmental Protection Agency
EU	European Union
EV	Electric Vehicle
GCC	Global Climate Change
GDP	Gross Domestic Product
GHG	Greenhouse Gas
GIB	Grid Independent Building
HDI	Human Development Index
HFC	Hydro-Fluoro-Carbon
IEA	International Energy Agency
IPCC	Intergovernmental Panel on Climate Change
ITER	International Thermonuclear Experimental Reactor
NDC	Nationally Determined Contribution
NIMBY	Not in My Back Yard
NOAA	National Oceanic and Atmospheric Administration
NZEB	Net-Zero Energy Building
OECD	Organization for Economic Cooperation and Development
PES	Primary Energy Source
PHS	Pumped-Hydro Storage
POF	Power Operating Factor
ppm	parts per million
ppb	parts per billion
PV	Photovoltaic
RES	Renewable Energy Source
RGGI	Regional Greenhouse Gas Initiative
SMR	Small Modular Reactor
UN	United Nations
UNFCCC	United Nations Framework Convention for Climate Change
US-DOE	United States Department of Energy
WCED	World Commission on Environment and Development

Preface

Human civilization is defined by the use of energy. Our civilization started when our prehistoric ancestors harnessed fire to light and heat their caves, and progressed with the use of animal power for transportation and food production. The Industrial Revolution with its energy-voracious engines, totally changed the human society. Mechanized agriculture liberated most of us from toiling the land for the production of our food. Factories use tremendous quantities of energy to supply furniture, goods, and gadgets to an increasing and more affluent global population. Homes with artificial lighting, refrigerators, heating, and air-conditioning are infinitely more hygienic than human dwellings of any other era. Fast means of transportation propelled by engines allow the humans of the 21st century to be in touch and take them to faraway lands. Entertainment is better than any other epoch, with every home having its own theater and music hall. The invention of engines that run on affordable energy has made possible all these amenities that have tremendously improved the human life and contributed to the development of nations.

This book starts with an introductory chapter that describes the several forms of energy and presents the enormous benefits of modern engines and energy usage to humans. After the Industrial Revolution and the energy-voracious thermal engines, the availability of sufficient and affordable energy has contributed to the development of nations, the prosperity of their citizens, and through prosperity, their citizens' better health and longevity. Historical data are presented on the global energy demand and supply, and their future trends are inferred from recent data and developments. A long section is devoted to electricity – the fuel of the 21st century – how it is generated and what are its future trends, given the insatiable demand for more electric power from datacenters, electric vehicles, and air-conditioning. The chapter concludes with an explanation of energy resources, energy reserves, and a detailed note on energy units and their conversions.

Having established the importance of reliable and affordable energy, including fossil fuels, the second chapter examines the warming effects of atmospheric carbon dioxide. The concepts of weather and climate are differentiated; data on the average global temperature and the atmospheric concentration of CO_2 spanning more than twelve decades are presented; and the cause-effect relationship between the two is explained, using several sets of data and graphs. The analysis is extended to include the effects of historical CO_2 emissions by the industrialized countries, as well as the contribution of a continuously increasing global population, primarily due to developing nations. Since the scientific community was successful in eliminating other environmental threats of the past – acid rain, atmospheric lead, and the ozone hole – the strategies that contributed to the elimination of these threats are presented, and the uniqueness of the current Global Climate Change threat is contrasted with the past successes. Because of the global notoriety of the climate change issues and the attention they have received from journalists, activists, politicians, and self-proclaimed experts, this chapter concludes with an exposition of several myths that were promulgated from all sides about the certainty of the warming effect as well as some outrageously catastrophic predictions.

The third chapter offers a succinct and critical survey of the technologically feasible methods for the control and eventual reduction of the anthropogenic CO_2 emissions. Among these, energy efficiency, higher use of nuclear energy, and transition to renewable energy – in particular wind and solar energy that are available everywhere on the planet – are the methods that appear to have a significant impact on CO_2 emissions reduction. These methods are also advocated in all the pathways and roadmaps to mitigate the Global Climate Change by the national and international organizations. Detailed data on the transition to renewables and the requirements to decarbonize at three levels – a single home, a city, and an entire grid – are also presented for the reader to appreciate the magnitude of the needed infrastructure. The chapter concludes with the consequences electric vehicles may have on decarbonization efforts and includes comments on the roadmaps presented by the agencies of the United Nations.

The fourth chapter probes into the economics of decarbonization and starts by explaining the volatility of prices and the uncertainty associated with any future cost predictions for energy systems and energy prices. The several types of energy subsidies and the carbon taxes are presented, and their effects are examined on the energy markets, the global electricity prices, and the poorer segments of the population. The relationship between the economics of decarbonization and the United Nations *Sustainable Development Goals* is critically examined, with particular emphasis on the Goals that are pertinent to energy use, climate change, energy transition, poverty, and inequalities within and among nations. The elements of the necessary infrastructure for a global decarbonization of the electricity industry are presented (including new solar and wind energy units, land procurement, energy storage systems, and electricity grid upgrades). The cost for building this infrastructure is calculated with reference to the electricity grid that serves Texas. It is concluded that the decarbonization investment will have a substantial effect on the electricity prices of the future, which will have to increase by a factor between 3.3 and 4.8, significantly affecting the poorer part of a nation's population and the entire nations of the developing world.

The enormous price for the decarbonization infrastructure and the future cost of the highly touted energy transition leads to an incongruity (conundrum) for the set of the four *Sustainable Development Goals* (SDGs) that are pertinent to poverty, inequality within and among countries, reliable and affordable energy, and combating climate change. With the firm conclusion that energy transition is incongruent with concurrent actions to eradicate poverty and reduce the inequalities within and among nations, the only realistic path for the future is the global adaptation to the effects of climate change and the prioritization of the SDGs, without abandoning the CO_2 emissions mitigation efforts. Providing the world's population with sufficient and affordable energy is the key to the SDGs prioritization strategy. The epilogue of this book concludes with observations on twelve trends on energy demand and supply that will endure in the climate adaptation era.

I am indebted to a number of individuals who helped with the conceptualization and writing of this book. First among them are the students who took my courses on *Power Generation*, *the Chernobyl Legacy*, *Alternative Energy Sources*, and *Sustainable Energy*, which I have offered in the last 46 years at four Universities (Delaware, Tulane, Texas at San Antonio, and Texas Christian). I have learned from them more than they have learned from me. I am very thankful to my several colleagues in these Universities as well as to other colleagues and friends I regularly meet at conferences. I have had many fruitful and sometimes animated discussions with them on energy, the environment, and the great energy challenge our society is facing. Kyra van Nostrand and Sonia Tam of the Taylor & Francis Group/CRC Press were always there to provide managerial support and advice. The production and publication team of CRC spent many hours putting together the chapters and sections of this book. I am also very much indebted to my own family, not only for their constant support, but also for lending a hand whenever this was needed. My wife, Laura, has been a constant source of inspiration and help. Our children, Emmanuel, Dimitri, Eleni, and Alexi, were always there and ready to help. I owe sincere gratitude to all.

Efstathios E. (Stathis) Michaelides
Fort Worth, Texas, May 2026

1 Global Energy Demand and Supply

The discovery and procurement of energy sources and the use of energy in its various forms have defined human civilization and have shaped the modern world. This introductory chapter describes the several forms of energy and presents the enormous benefits of the modern engines and energy usage to all humans. Following the Industrial Revolution, the availability of energy sources has contributed to the development of nations, their affluence, and through affluence their citizen's health and longevity. Historical data are presented on the global energy demand and supply and their future trends are inferred from recent data. A separate section is devoted to electricity – the fuel of the 21st century – how it is generated and what are its future trends given the voracious demand for more electric power from data centers, electric vehicles, and air-conditioning. The chapter concludes with an explanation of energy resources, energy reserves, and a detailed note on energy units and their conversions.

1.1 ENERGY AND HUMAN PROGRESS

The use of energy has defined human society and human civilization. When the prehistoric humans mastered the use of fire to light and heat their caves (what we now call "domestic comfort") and to cook their food, human civilization began. Centuries later, the use of animal power (primarily horses and oxen) significantly contributed to the development of ancient and medieval agricultural societies and helped provide food to nourish an increasing human population. More recently, the Industrial Revolution and the invention of energy-voracious engines created today's knowledge-based society, where citizens live in homes of greater comfort; increasingly use fast modes of transportation for travel; and are continuously in contact with modern communication devices. In the last two centuries the human society has evolved by using increasingly large quantities of energy and humans have arrived at the point where they cannot survive without it. The consumption of vast quantities of energy – in thermal, chemical, mechanical, and electrical form – is absolutely necessary for the functioning of the contemporary society, the prosperity of nations, and the survival of our civilization.

While we commonly use the word "energy" for several related concepts, all of us use energy *per se* in a variety of different forms for lighting, heating, cooling, communications, and automotive power. Complex global networks supply sufficient power to ensure that our demand for energy is satisfied any time we need it. Airplanes, ships, and automobiles use liquid hydrocarbon fuels derived from crude oil; coal, in its many forms, is used for the production of steel and cement; many contemporary households use natural gas for cooking and heating, and electricity for lighting and entertainment; computers and wireless network systems use electricity. Yet, electricity is not found

DOI: 10.1201/9781003674658-1

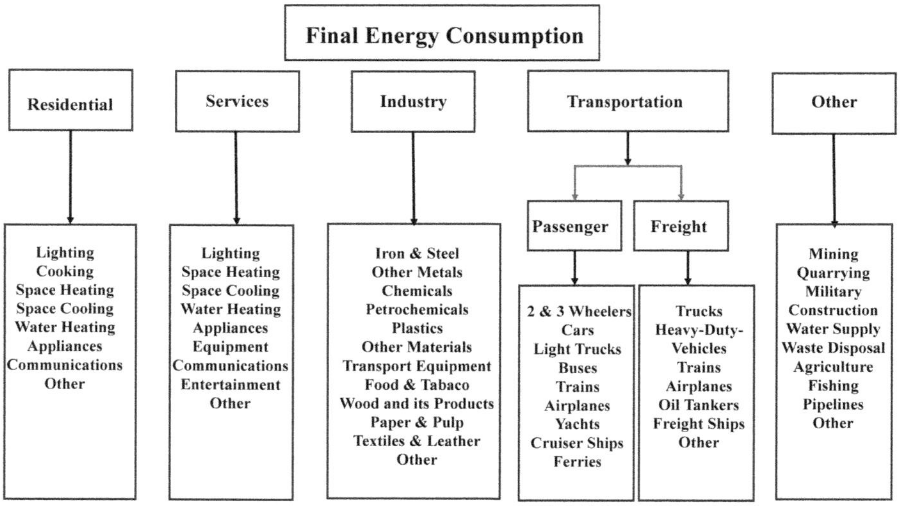

FIGURE 1.1 Principal energy consuming sectors in the modern society.

on earth as a natural resource. It is generated in a variety of power plants that convert *primary energy sources* (PESs) that naturally exist on earth – e.g., coal, natural gas, nuclear, solar, wind, geothermal, hydroelectric, etc. – to electric power.

The modern human society has developed elaborate and technologically advanced networks that transport and supply the needed energy to the final consumer. Electricity is fed into communities by the transmission lines of the electric grid at high voltage. Natural gas is transported to households by a complex system of pipelines, which often transcend national boundaries. Tanker ships crisscross the oceans daily to bring crude oil to refineries, which in turn supply us with gasoline and diesel fuel via an intricate system of pipelines, train cars, and trucks. Figure 1.1 shows the sectors where energy is consumed in the modern global economy. From cooking at home to mining rare earth metals, energy is consumed in large quantities by an ever-increasing global population.

In today's world, energy in its different forms defines the life of contemporary humans, and enters all aspects of human life, economics, and politics. The impact of energy supply and the trade of energy is vital to the wellbeing and the economies of all nations. The geopolitical activities of nations have been significantly influenced by the need for an adequate, reliable, uninterrupted, and secure energy supply. Several wars, since the beginning of the 20th century, have been fought for the control and security of energy supplies. Treaties, alliances, and international agreements have been cemented with energy resources as the primary issue.

1.1.1 What Is Energy?

The word energy (ενέργεια) was coined by Aristotle, to denote the inherent ability of objects to perform work. It is a composite word made of the preposition εν, which means in or inside and the word έργον, which means work. An operational translation of the word ενέργεια would be "work content," or the capacity of materials and systems to perform work. Everything has the potential to do something, and ενέργεια is the capacity or the manifestation of that potential to act. The modern science of thermodynamics uses the word in a similar manner, with several engineering textbooks defining the property *energy* of the materials as their "potential to perform work."

Energy in materials comes in different forms: potential energy (related to gravity), kinetic energy (related to velocity), elastic energy, magnetic energy, electric energy, surface energy, chemical energy, thermal energy (heat), nuclear energy (mass of radionuclides), etc. All these forms of energy can be transformed by pieces of machinery to perform tasks for humans such as lifting weights, compressing and transporting gases and liquids, propelling vehicles and ships, heating houses, and so on.

The existence of all these energy forms can be confusing. Richard Feynman, one of the most famous physicists of the 20th century, added to the confusion by stating that "... we have no knowledge what energy *is* .... There are formulas for calculating some numerical quantity, and when we add it all together ... (we get) always the same number" [1]. I sometimes hear the opinion that since we do not know what energy *is,* we should refrain from doing much with it until we fully grasp its essence. As an engineer, I do not subscribe to this notion, which often leads to an intellectual nihilism summarized by the Socratic paradox *έν οίδα ότι ουδέν οίδα* (*I know one thing that I know nothing*). There are many forms of electromagnetic waves, and whether we "know" what they are or not, we use them every day in radios, televisions, cell phones, microwave ovens, and radiation heaters. Similarly, there are many forms of potatoes, we may not know what they *are* exactly, but they are nutritious, and we eat them daily. Engineers have built bridges, airplanes, microwave ovens, telephones, and computers using an empirical knowledge of their parts and the energy that enables them to function and serve us. Energy in this book will be treated in the same empirical or heuristic, and universal manner. In our own ways we all know what energy *is,* how much of it we use, and how much we pay for it.

1.1.2 Life Was Not Fun without Energy – the Transformation of the Global Community

The most important benefit of energy is to liberate the human race from working on the land to produce their food. I have read several books (both historical and fictional) and often watch films relevant to the medieval times and talk with my children and students about similar shows, such as *The Game of Thrones*, which was a global sensation in the 2010s. What is always missing in these epics is the representation and depiction of the vast numbers of serfs and peasants, who daily worked the land and provided food for the illustrious protagonists – kings and queens, princes and princesses, dukes, knights, and all the other notables in the shows. Those who worked on the land to feed everyone made up for an overwhelmingly high fraction of the global population. In the high Middle Ages, close to 92% of the European population were tied to the land (many as serfs) with the task of working 24/7 to produce food for the "nobbles," and for the entire community to survive. At the same time – because of lack of knowledge, hygiene, and what we now call "health providers" – the average person in Europe and Mesoamerica lived for close to 30 years. Life was not as glamorous in the age of dukes and knights as some of the modern films would have us believe. I often remark in my classes (to a captive audience of 20-year-olds) that if we all lived in the thrilling age of the thrones, the vast majority of us would have been illiterate old people working the fields every day of our lives.

This rather bleak picture of human society drastically changed with the Industrial Revolution and the appearance of energy-voracious engines and energy-consuming chemicals. Tractors, fertilizers, pesticides, and combines, transformed the agricultural societies and liberated most humans from having to work for the sustenance of their communities. While in the 1420s farmers produced a little more food than what was enough to feed their family, in the 2020s a farmer in Kansas produces enough food for another 128 people! A fleet of vehicles (ships, trains, trucks) brings the food to the markets of the cities and to the tables of modern humans. And the 128 people, who do not have to work the land anymore, are free to pursue their own careers as engineers, accountants, doctors, nurses, drivers, electricians, teachers, and entertainers. The use of energy, in its different forms, has liberated the human race from toiling the earth to produce food. They may now pursue other professions and more pleasant activities (including writing books on sustainability!) that fulfill their lives; follow their personal goals to find happiness; travel for vacations; and be more affluent than the humans of any other epoch.

When visiting older buildings dated before the Industrial Revolution, I am always astonished at the lack of amenities and the sparse furnishings. Even the households of the mighty and affluent (from the Versailles Palace, in France, to Monticello, in Virginia) lack comfortable sofas, armchairs, coffee tables, and spacious desks. Of course, refrigerators, electric coffee makers, and dishwashers are nowhere to be seen. The human dwellings of previous eras were sparsely furnished, badly ventilated, unevenly heated in the winter, and never cooled in the summer. Using a variety of modern tools and machinery, the factory workers of the 21st century produce by far more chairs, desks, sofas, and tables than their colleagues of a few centuries ago. Energy-voracious engines also supply the modern household appliances that run on electric power and have made our lives comfortable. The vastly increased supply of consumer products has made furniture and appliances affordable for the average household and has enabled most of us to have comfortable homes and places of work – even though we now have twelve times more households than before the Industrial Revolution. Heating, ventilation, and air-conditioning (A/C) units are ubiquitous and have made modern dwellings to be much better and infinitely more hygienic than the dwellings of the past. The availability and meticulous training of growing numbers of doctors, nurses, and other health professionals have enabled the global population to increase eightfold and for the life expectancy to continuously reach new heights. New and faster means of transportation – airplanes, cars, ships, and bullet trains – bring people, families, and nations together more often than in previous centuries. Modern computers, a multitude of new communication devices, and an electronic web that covers the wide world disseminate sounds, images, videos, and information throughout the globe instantly. High productivity, which is made possible by engines that consume ever-increasing quantities of energy, and ease of travel have allowed the modern humans to take vacations (a practice totally unknown to the general population a few centuries ago) and visit far-away places. The invention of engines and the availability of affordable energy has made feasible all these amenities that have improved the human life.

Labor, capital, entrepreneurship, and land – four fundamental inputs in the economies of nations – are considered as the primary factors of production in economic theory. Energy is an implicit factor among them because it transforms resources (materials) into useful products and has become the indispensable link between labor and capital. In the modern world the capital provides the machinery, while the labor operates the engines using energy. Without energy the engines do not work and the labor-capital collaboration, which has ushered in the economic advancements of the last two centuries, breaks down.

In short, the consumption of energy in a multitude of complex engines and systems has liberated the humans from working on the land for their food; and enabled them to choose careers outside of farming. As a result, the humans of the 21st century are more affluent; live in comfortable dwellings; live healthier lives; live longer; communicate instantly with the rest of the world; and travel throughout the world for business and pleasure. All this progress has become possible because the humans of the 21st century use 700 times more energy than their ancestors of the 18th and previous centuries.

1.2 ENERGY AND THE DEVELOPMENT OF NATIONS

The prosperity of a nation and the affluence of its citizens are directly related to the amount of energy the nation consumes. Citizens of the more affluent nations buy more consumer goods, which are energy-intensive; frequently use private means for commuting to their work, such as cars (and, the extremely affluent, private helicopters); travel often and further for vacations; make use of a myriad of energy-consuming appliances; and have higher levels of energy-intensive comfort in their homes, like central heating and A/C. All these amenities consume large quantities of energy. On the contrary, the citizenry of the less affluent nations, on the average, use a great deal less of these engines, appliances, and systems. As a result, they consume significantly less energy. When the economies of developing nations improve, their citizens buy more electric appliances for their homes; commute in private vehicles; install A/C systems at home; and travel more frequently. The proliferation of such activities substantially increases the energy consumption of the developing nations. Figure 1.2 depicts the gross domestic product (GDP) *per capita* versus the primary energy consumption *per capita* (in a logarithmic scale) of several representative nations [2]. The GDP values are in 2023-US$ and were adjusted to take into account the different purchasing power of the US$ in the respective countries [3]. The figure shows the relationship between societal affluence and energy use. Even though there is a good measure of scatter in the data, it can be easily seen that there is a convincing correlation between affluence and energy consumption: for every $4,000 to $6,000 increase of the GDP *per capita*, there is a corresponding increase of 1 ton of oil equivalent (toe). When nations become more affluent, they slide toward the top and the right side of the figure, approximately following the trendline. A prime example of this phenomenon is the development of the People's Republic of China,

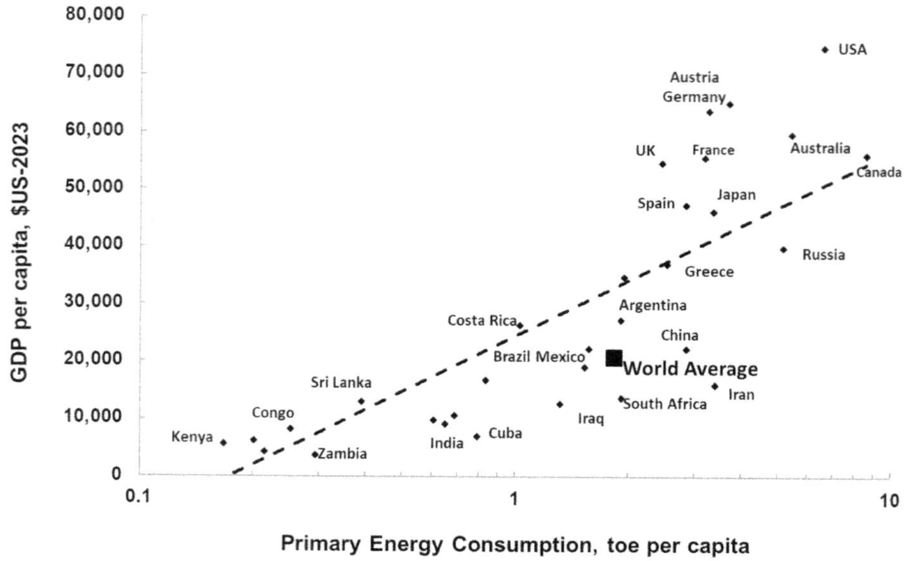

FIGURE 1.2 Gross domestic product *per capita* and primary energy consumption (in logarithmic scale).

whose GDP increased from $1,146 to $22,138 *per capita*, while its energy consumption increased from 0.31 to 2.86 toe *per capita* during the 40-year span from 1983 to 2023.

It is apparent in Figure 1.2 that there is a direct relationship between the perceived affluence of a nation and the average energy its citizens consume. Nation members of the Organization for Economic Cooperation and Development (OECD) [4], which encompasses most of the industrialized (and more affluent) nations have significantly higher GDP *per capita* and consume substantially more energy than developing nations. The OECD countries appear at the upper-right part of Figure 1.2. At the other end of the figure are developing countries where the average citizen is less affluent and the energy consumption is substantially less. In the middle of the graph, there are nations in intermediate stages of economic development. The energy consumption in these countries is commensurate with their respective GDP. Among these middle-tier economies the countries of Brazil, Russia, India, China, and South Africa (sometimes referred to as "BRICS") have significantly improved their GDP in the period 2000–2025, and this is reflected in their energy consumption. The citizens of these, and other similarly developing countries, have become more affluent in the 21st century. As a result, the energy consumption in the five nations has significantly increased.

The Human Development Index (HDI) is a quantitative measure of the quality of life in the United Nations (UN) member states. It was developed by the United Nations Human Development (UNHD) program of the UN [5]. The index was introduced in 1990 to emphasize that it is the people of a nation and their capabilities and not economic growth alone that define the development of nations. The HDI, which is a number between 0 and 1, is a statistical measure that takes into account quality

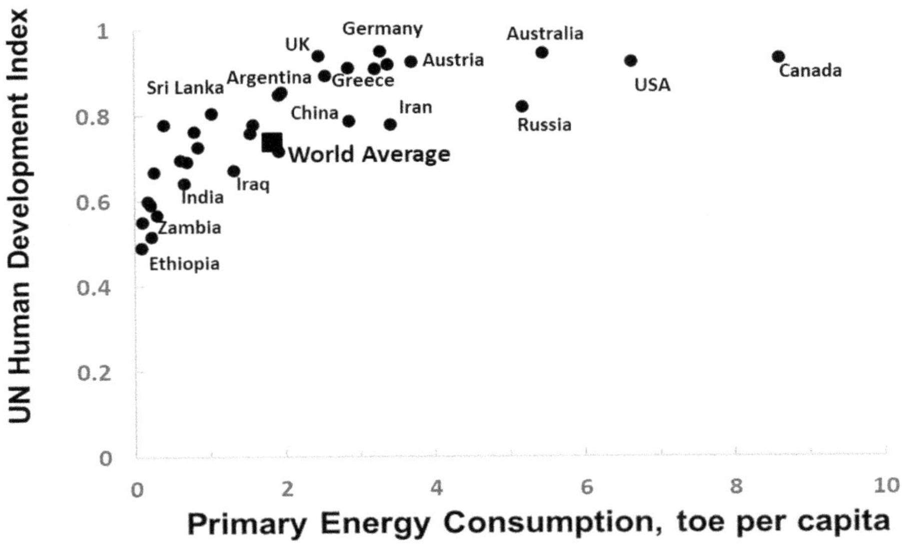

FIGURE 1.3 Human development index and primary energy consumption.

of life indicators for each nation such as: the wellness and health of the citizens (as measured by the life expectancy at birth), public education and years of compulsory schooling (for both males and females), population growth and fertility rates, rate of inflation, income *per capita* adjusted for purchasing power and income inequality among the citizens [6]. Figure 1.3 depicts the relationship between the HDI and the total primary energy consumption *per capita.*

One can easily observe in Figure 1.3 that there is a significant correlation between the HDI (the ultimate criterion for assessing the development of a country according to the UN) and energy consumption. Countries have achieved high levels of development (as measured by their HDI) primarily because their economies grew using increasingly large quantities of energy; they are currently more affluent; and they are still consuming high quantities of energy *per capita* to satisfy their needs.

It must be noted that in both Figures 1.2 and 1.3, there are a few countries, most notably the USA and Canada, that are outliers in the graphs because they consume too much energy. This is primarily due to the culture of long commutes in private cars and the predominant single-family residences that require a great deal of energy for heating in the winter and cooling in the summer. Undoubtedly, the USA, Canada, and a few other developed nations can adopt energy conservation measures that would reduce their energy consumption. This fact notwithstanding, the message from the data of Figures 1.2 and 1.3 is clear: economic prosperity and high levels of human development are intertwined with the consumption of substantial amounts of energy.

One of the components that comes into the calculation of HDI is citizens' longevity, as measured by the average life expectancy at birth. Figure 1.4 includes the data on life expectancy versus primary energy consumption in the same countries that are

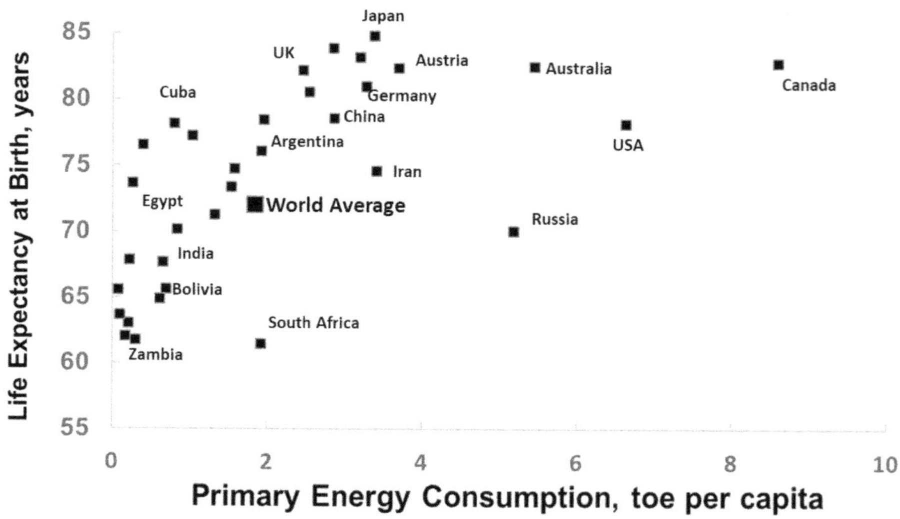

FIGURE 1.4 Life expectancy at birth and primary energy consumption in representative countries.

depicted in the previous two figures [7]. I am always amused to point to my students the correlation between nations of healthy and longevous citizens (on the average) and the quantity of energy they consume. Since I am categorized as a senior citizen, and 15–18 years make a big difference in my life horizon, I am happy that I have lived my life in countries that are in the upper part of the figure. Regarding longevity, it is important to note that energy consumption is not the *cause* of longevity in the 21st century, but the *effect* of prosperity and all the implicit benefits associated with it, which require higher energy consumption, including prenatal care, vaccine availability, better public health, availability of trained health care practitioners, healthy dwellings, and a healthy environment.

Anthropologists and historians have established a cause-and-effect relationship between longevity and energy consumption by examining human remains and energy use since the beginning of human life. Prehistoric humans had to perform all the tasks by themselves and lived in unsanitary caves. The fossils of their bones reveal a great deal of "wear and tear" for their bodies, and modern scientists estimate that their life expectancy was close to 20 years. In the early historic period humans domesticated animals, such as horses and oxen, to perform the most arduous tasks, but still worked hard in agriculture and did not live in healthy residences. The life expectancy of these humans increased to about 30 years. The humans of the 21st century are employing machinery to do most of their manual work including manufacturing and agriculture; their vast majority do not physically stress their bodies (unless they do it for sport or recreation); live in well-ventilated, heated, and cooled homes; and enjoy private and public health care. All these characteristics of our historical era require the consumption of a great deal of energy and, as a result, the human life expectancy has increased dramatically in the last two centuries. While it is often said that "money does not buy happiness," Figures 1.2 and 1.4 demonstrate that, at least, affluence with its associated high energy consumption can buy a longer life.

1.3 GLOBAL ENERGY DEMAND AND SUPPLY

A glance at Figure 1.1 proves that energy usage pervades all sectors of the national economies, it actually drives the business and the everyday life of citizens. Energy does not come in a single form but in several forms, which are promptly consumed in increasing quantities by a growing and more affluent global population. Electricity is used for lighting, motors, heat pumps, and A/C units for domestic comfort, computers, data centers, and more recently, for electric vehicles (EVs). Petroleum products (gasoline, diesel, kerosine, etc.) are primarily used for transportation. Traditional biomass (wood from trees and bushes, and agricultural waste) is still used in several parts of the developing world for cooking and heating. Coal, in its various forms, and natural gas are used in many industrial processes to generate heat, and also for electric power generation. Other energy forms, which are primarily used for electric power generation are nuclear, hydraulic, solar, geothermal, and wind energy.

In addition, energy resources are consumed in "non-energy uses" for the production of useful materials. Plastics are manufactured from petroleum distillation products; natural gas is converted to agricultural fertilizers; anthracite is essential for the production of steel; and cosmetics, ointments, and lubricants are manufactured using materials that become available in the petroleum distillation process.

Most of the useful energy forms and non-energy products do not exist in nature. They are produced from natural resources, the *primary energy sources* (PESs). These resources exist in nature and can be mined or harnessed in the natural environment. The following are the three main categories of PESs:

1. Fossil fuels are the various forms of coal (anthracite, lignite, bituminous, peat, etc.), petroleum (crude oil), and natural gas.
2. Nuclear fuels, primarily uranium and thorium, are contained in naturally occurring minerals, such as uranite, carnotite, autunite, monazite, thorianite, and thorite.
3. Renewable energy forms such as hydroelectric, solar, wind, biomass, waste, geothermal, tidal, and wave energy.

These natural energy resources are not uniformly distributed on the planet. The Arabian peninsula is endowed with a great deal of petroleum; the subterranean parts of the USA, China, the UK, and Germany contain large quantities of coal; large deposits of uranium and thorium exist under the Sahara in Mali and under the mountains of Canada. Even the renewable energy sources have preferential regions. Sri Lanka, Morocco, and Sudan enjoy substantially more solar irradiance than France. A great deal of geothermal energy can be derived in Iceland and the Central American countries, but not in Austria. Within Nepal and Norway there are rivers with high potential to produce hydroelectric energy, but Australia and Vietnam have few such resources. The winds in Denmark and North Texas are strong enough to generate significant quantities of electricity, but winds are weaker in Brazil and India.

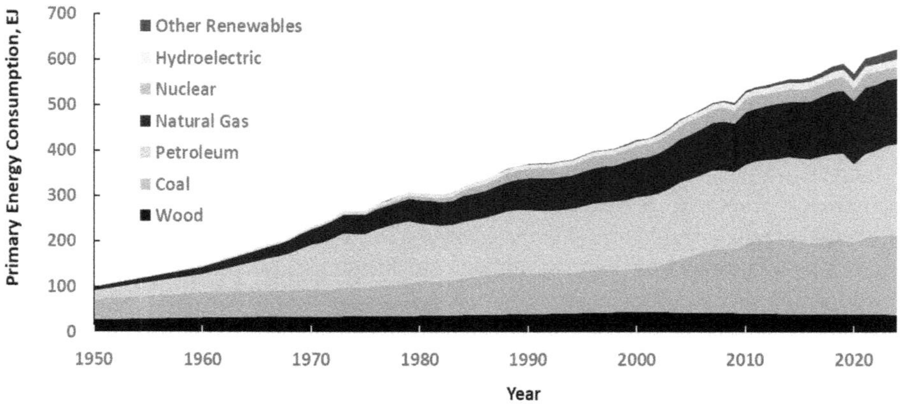

FIGURE 1.5 Global primary energy supply between 1950 and 2024. From bottom to top: Wood, coal, petroleum, natural gas, nuclear, hydroelectric, and other renewables (biofuels, solar, wind, geothermal, tidal). 1 EJ = 23.9 million toe.

Because of this non-uniformity of the energy sources, a vigorous energy trade has been developed among the nations of the globe: petroleum flows from the Arabian peninsula to Europe, India and the Far East; natural gas is transported in pipelines from the Russian Federation to Europe and China; and wind-generated electricity is transmitted in high-voltage cables from Denmark to Germany and from France to Spain.

Figure 1.5 depicts the total global supply of PESs from 1950 to 2024 [8]. Non-energy uses of these natural resources are not included in the figure. One may observe that the total global primary energy supply increased by a factor of 6.4 (740%) since 1950 (when the automobile became a household item in the OECD countries) and keeps increasing at a robust pace, following the trends of an increasing and more affluent global population.

One can easily see in Figure 1.5 that the consumption of traditional biomass (primarily wood) has been almost constant in the 75 years from 1950 to 2025, at approximately 40 EJ [8]. The consumption of all types of coal was steadily increasing until the early 21st century, it is increasing at a slower pace since 2015 and has reached a level of approximately 170 EJ. This happens despite all the recommendations from national and international environmental organizations to reduce the use of coal that produces a great deal of CO_2. The consumption of petroleum has been steadily increasing because of the fast-growing number of automobiles in the developing world in the recent years. Similarly, the use of natural gas has been increasing, following the technological advances in natural gas wells (fracking and horizontal drilling) that increase its supply and reduce its price, and the development of high-efficiency gas turbines for the generation of electric power. The global contribution of nuclear energy has been almost steady in the last thirty years primarily because of risk concerns (after the accidents in Three-Mile Island, Chernobyl, and Fukushima). The generation of hydroelectric energy is increasing at a relatively slow pace. It is also apparent in Figure 1.5 that the use of the other renewable energy sources (primarily biofuels other than tree-wood, solar, wind, and geothermal) is rapidly growing, following the significant investment of several nations in these energy sources. However, the fraction of the non-carbon energy sources – renewables including wood, and nuclear – is still very low. In 2025, the combined contribution of these PESs was less than 17% of the total.

1.3.1 Recent Trends and Projections

A glance at Figure 1.5 proves that the global PES consumption has been increasing. A closer analysis of the data shows that, since the beginning of the 21st century, the growth of PES consumption slowed – from an annual average of 2.9% in the late 20th century to 1.6% in the 21st century. However, the slower rate of growth is not projected to continue in the future. The publication of the 2025 *Global Energy Review* by the International Energy Agency (IEA) alerted the global community that the growth of PESs rose substantially in 2024 (to 2.2%) and very likely will continue growing at a higher rate in the next decade [9]. About 80% of this growth is attributed to developing economies. A significant finding in the report is that, after several years of energy consumption decline, the PES consumption in the advanced economies grew by almost 1%, primarily due to higher electricity demand. The growth of the global electricity demand is expected to persist in the coming decades (perhaps at an accelerated pace), and, for this reason, it is expected that the global demand for PESs will continue at a similar pace.

It must be noted that, even though renewable energy sources accounted for a large share of the growth of the global energy supply in the 21st century; and that renewables have met approximately two-thirds of the global electricity growth in 2024; the global consumption of fossil fuels – natural gas, coal, and petroleum – also grew significantly with no signs of decline. Fossil fuels still account for more than 80% of the total PESs globally consumed in 2024 and 2025 [8, 9]. Based on the data of the first quartile of the 21st century, it is highly unlikely that the continuously growing and more affluent global population will curb its appetite for energy, including fossil fuels.

1.3.2 Energy and Carbon Dioxide Emissions

Global Climate Change (GCC, or global warming for short) is the most pressing environmental issue of the 21st century that threatens not only the global economies but also the ecosystems on the planet. While global warming is examined in more detail in Chapter 2, it is important to note which PESs contribute most to the atmospheric CO_2 emissions. Coal, in its various forms, is the number one culprit. Coal contains primarily carbon and its combination with the atmospheric oxygen (combustion) produces CO_2. Petroleum and natural gas are fossil fuels made of hydrocarbons. Their molecules are made of chains of carbon atoms connected with a bunch of hydrogen atoms. When they burn, the carbon part produces CO_2 and the hydrogen part water, H_2O – a benign to the environment substance. Natural gas is composed of light hydrocarbons, primarily methane, CH_4. The combustion of fossil fuels supplies heat, which is needed for the generation of electricity, internal combustion engines for vehicles, the production of cement and concrete, and a myriad of other applications. When fossil fuels are burned, coal produces substantially more CO_2 per unit heat delivered than the other two fossil fuels. Natural gas produces the least amount of CO_2 per unit heat delivered, and the petroleum products (diesel, gasoline, etc.) are somewhere in the middle.

Biomass is considered a *carbon neutral* fuel. The carbon in the biomass molecules is formed from the absorption of atmospheric CO_2 during the photosynthesis process of plants. Even though biomass produces CO_2 when burned, this gas has been extracted from the atmosphere. Consequently, burning biomass is a zero-sum game and does not add to the atmospheric concentration of CO_2. It must be noted, however, that actively growing biomass to produce fuels (e.g., corn to produce ethanol) is an energy-voracious and particularly inefficient way to produce fuels because it consumes more primary energy forms than the energy it finally delivers [10, 11]. Growing biomass as an energy source may offer a great deal of benefits to farmers, but it is not a viable answer to our energy challenge.

Nuclear and the renewable energy sources – solar, wind, geothermal, hydraulic, tidal, etc. – are considered CO_2-free sources. Except for geothermal energy, which may release traces of CO_2 into the atmosphere, these energy sources deliver heat and electricity without emitting this gas. The machinery and equipment used for the harnessing of the energy in these sources may have a "carbon footprint," but their operations to generate useful heat and electric power are largely clean.

Looking back at Figure 1.5, one can see that, despite the several pleas and recommendations by the United Nations agencies and the corresponding national plans, the use (and combustion) of fossil fuels has been steadily increasing. The atmospheric CO_2 emissions are also growing at a commensurate rate. Even though we have recently seen a robust growth in renewable energy units – primarily solar farms and wind farms for electricity generation – the contribution of fossil fuels to the global energy consumption is stubbornly hovering around 80% since the 1990s, when GCC became a global environmental priority. As a result, the CO_2 emissions and the atmospheric concentration of this gas keep increasing.

1.3.3 Can We Live with Less Energy?

Since the incessant use of more energy and fossil fuels has become a major environmental threat, one often wonders if we can use less energy and, thus, curb the atmospheric CO_2 emissions. Looking again at Figure 1.5, it is apparent that there have been several years when the global energy consumption has dropped rather than increased. The downward bumps in Figure 1.5 deserve a closer examination and this is done in Figure 1.6, which depicts the annual increases or decreases of the global energy consumption in percentages. The labels at the troughs of the graph describe the geopolitical events that are associated with the drop in the global energy use. From the energy crisis in the mid-1970s to the Covid-19 pandemic in 2020, all energy consumption reductions are associated with significant global economic downturns that have been harmful to the global citizenry. In the case of the most recent such occurrence, no one wishes a repetition of the energy reduction of 2020 (and its associated CO_2 emissions reduction) if this implies the global suffering and economic hardships related to the pandemic of Covid-19.

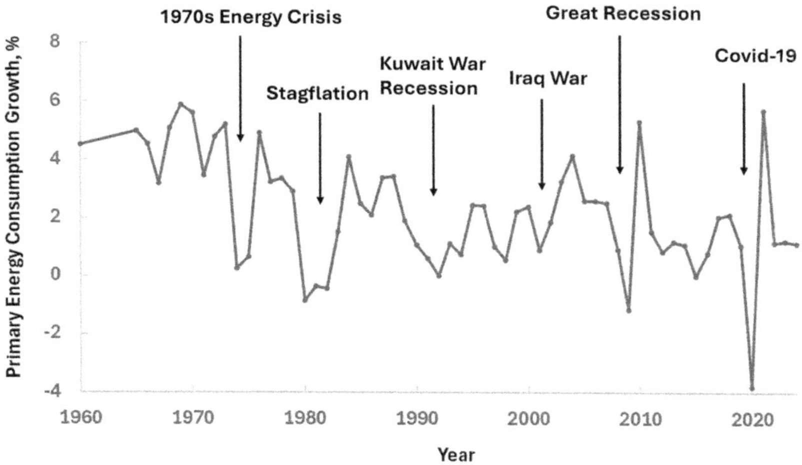

FIGURE 1.6 Annual fluctuations of global energy consumption in %. The labels show the associated geopolitical events.

A related question is: can the world economy function well with less energy? The short answer to this question is that yes, the global economy can function well with less energy consumption if conservation and higher efficiency measures are widely adopted. Installing a highly efficient A/C unit (e.g., one that dumps heat in a cooler pond or the cooler underground soil rather than the hot surrounding air) in the summer or asking the family to put on an extra sweater and keep the thermostat at 18°C (65°F) in the winter are actions that reduce energy consumption at home. Developing vehicle engines that deliver higher mileage rates (and favoring smaller vehicles!) would decrease the global crude oil consumption. However, some of these actions are costly (e.g., the high-efficiency A/C unit) or unpopular (the low temperature at home) and are not likely to be voluntarily adopted. Governments that intensely regulate such stringent efficiency and conservation measures risk becoming unpopular and will not strongly push the issues with a fickle electorate. In short, energy conservation and higher efficiency will somehow reduce the global energy consumption but cannot be relied to substantially curb the combustion of fossil fuels.

1.4 ELECTRICITY – THE FUEL OF THE 21st CENTURY

Electricity is the most convenient energy form for consumers. We turn on a switch and good things happen: lights in the house, entertainment on demand, heating or cooling at will, instant communications, non-polluting urban subways, fast bullet trains, etc. The developed nations have benefited from the use of large quantities of electricity for decades and the developing world is now catching up. The growth of electricity demand has always been higher than the growth of primary energy and hovered in the range 2.3–2.9% annually. However, in the middle of the third decade of the 21st century, it appears that this growth rate is accelerating. In February 2024, the IEA announced [12] that the predicted global electricity growth in 2024 would be 3.4%, a significant departure from the norm. One year later we were informed by the same agency that the actual global electricity growth rate for 2024 was a whopping 4.3% (1080 TWh) [9]. What is more remarkable with these figures is that the acceleration of the rate of electricity consumption is not restricted to the developing world but includes the advanced economies, which accounted for 230 TWh or 21.3% of that growth. Even the European Union (EU), which, for years, had put a lid on the consumption of electricity, with almost 0% growth (negative in several years), experienced a significant increase of 1.5%.

There are four principal reasons for the recently accelerating global electricity demand:

1. The increasing use of A/C systems for cooling in the developing world.
2. The dawn of the artificial intelligence era that necessitates electricity-voracious data centers in the developed nations.
3. The continuing global proliferation of EVs.
4. The worldwide increase in industrial production and consumer goods, driven by an increasing and more affluent global population.

The world is entering a new era of industrialization with electricity becoming fast the dominant fuel of the economies in the 21st century. Figure 1.7 shows the historical growth of the global electricity demand since 1990 together with the PESs that were consumed to generate this demand. The small bump in 2024 (a growth trend that is expected to continue) is obvious in this figure.

Electric energy is not found in nature as coal, natural gas, wind energy, and solar irradiance are. It is a *tertiary energy form* and needs to be generated from PESs, such as coal, nuclear, solar, wind, etc. It is a consequence of the 2nd law of thermodynamics that the primary energy spent on the generation of electric energy from

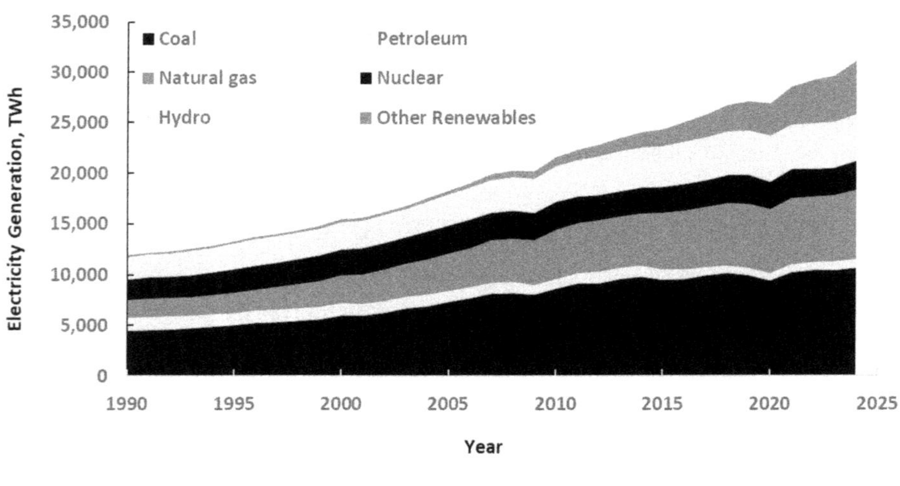

FIGURE 1.7 PESs spent on the generation of electricity in the period 1990–2024 (note on units: 1 TWh = 0.0036 EJ).

the PESs is substantially higher than the energy recovered from the conversion [13]. According to the IEA, the world generated 31,153 TWh (equivalent to 112.2 EJ) of electricity in 2024. The energy quantity of the PESs consumed for the generation of this electricity by far exceeds the 112.2 EJ value. Approximately 68% of the global electric energy was generated in thermal power plants – fueled by coal, petroleum products, natural gas, and nuclear energy – that convert heat to electric power and consume substantially more units of heat than the electricity they generate. A measure of this difference is the efficiency of the thermal power plant – the ratio of the generated electricity to the heat consumed. The efficiencies of the best coal power plants are close to 40%, implying that these plants consume 2.5 units of heat to generate one unit of electricity. Most nuclear power plants have efficiencies close to 30%, and the efficiencies of modern natural gas plants (combined cycle) can be in the range 50–60%. In total, the thermal plants alone generated 21,660 TWh of electricity (equivalent to 78 EJ) in 2024. These thermal power plants consumed approximately 211 EJ of PESs, mainly coal, nuclear, and natural gas, for an average efficiency of 37%.

Referring to Figure 1.5, a great deal of the primary energy consumption attributed to coal and natural gas, as well as a small fraction of petroleum is actually spent on the generation of electric energy. Almost all of the nuclear energy in that figure is used for electricity generation. Most renewable energy – photovoltaic (PV) solar, wind, and hydraulic energy – is directly converted to electricity, without the intermediate heat generation. It must be noted that – although there are finite efficiencies for the conversion of the renewable RESs to electricity – the numbers reported in Figures 1.5 and 1.7 reflect the quantity of generated electricity and not the solar irradiance absorbed in PVs, the available wind energy, and the potential energy of the water that have been used for the generation of electric power.

1.4.1 Waste Heat

One may reasonably ask what happens to the difference between the thermal energy input and the electric energy output of the thermal power plants. It is a consequence of the 1st law of thermodynamics that energy is neither created nor destroyed. It is transformed from one form to another, including heat (thermal energy). The thermal power plants consume a great deal of heat to generate electricity and the thermal energy that is not converted to electricity is discarded into the environment (the air, a river, or a lake) as waste heat. In our cars, this waste heat is discharged as the hot gases in the exhaust. A nuclear power plant that generates 2,000 MW of electric power and has an efficiency of 32% consumes 6,250 MW of heat from the nuclear fuel and discharges in its environment approximately 4,250 MW of heat. According to the numbers in the last section, the global production of waste heat in 2024 was $211 - 78 = 133$ EJ, by all means an extremely large rate of thermal energy!

The problem with the further utilization of waste heat is that it is released in the power plants, which are typically located outside urban areas, at very low temperatures, in the range 32–40°C (90–104°F). These temperatures are too low for the heat to be economically conveyed by a fluid (water) in long pipelines and be used in processes such as, for example, hot water or heating for the buildings of a city, a method called district heating. The low-temperature waste heat is a hot topic for non-scientists, who often advocate that these vast quantities of heat energy could be used to generate more electric power or in other conservation projects. There is not a month passing by, when I do not receive an email or a telephone call from a prospective "inventor" who claims to have "designed" an engine to utilize the waste heat of thermal power plants and generate substantially more electricity. Most of these "inventors" want me to find them a team of students (free labor!) who will build their "engine" that will save the world. Invariably I politely explain the science to these "inventors" and refuse to engage my students in such hopeless projects.

Projects that may use a small fraction of the waste heat are possible where low-temperature heat is required (e.g., aquaculture in warmer ponds, heating of greenhouses, subsurface heating of nearby fields to increase the yield of agricultural crops, etc.) are possible. However, such uses of waste heat are few and use a very small fraction of the waste heat produced globally. In 2025 less than 0.5% of the waste heat globally produced by thermal power plants was utilized, primarily because it is delivered at a low temperature, and has very little value. As Professor Smil would have put it, … this is how the world really works [14].

Another misconception that surrounds the waste heat is that, by increasing the temperature of the local environment, it significantly contributes to the global warming. With an average thermal efficiency of 37%, the thermal power plants also rejected approximately 133 EJ of waste heat in the biosphere in 2024. In the same year the earth received 5.46 billion EJ of solar energy or 41 million times more than the waste heat of the power plants. Even when we account for the heat produced by the engines of cars, trucks, ships, and airplanes, we will conclude that the waste heat produced by all the anthropogenic activities is several orders of magnitude less than the heat trapped in the biosphere by the greenhouse gases. It is the CO_2 gas, which is produced by the combustion processes and released in the atmosphere, that causes global warming, not the waste heat of power plants and the exhaust gases of the cars.

1.4.2 Trends and Projections for Electricity Generation

A glance at Figure 1.7 leads us to the following conclusions:

1. The global electric energy demand is constantly increasing. If the recent growth rates of close to 4% persist, the global electricity demand will nearly double by 2040.
2. A great deal of the global electricity growth is met by the fast-growing renewable energy units, especially wind and solar. The use of hydroelectric energy is also increasing steadily, but not as much as wind and solar. It is currently supplying close to 15% of the globally generated electricity, slightly lower than the sum of wind and solar electricity supply.
3. Coal – the most active contributor to the atmospheric CO_2 emissions – is still the major PES for electricity generation with 34% share. It is important to note that the quantity of coal burned for the generation of electric energy is also increasing, despite the Paris Agreement of 2015 and numerous appeals by environmental groups for its drastic reduction.

4. The contribution of nuclear energy to electricity generation has remained almost constant since 1990, and hovers in the range 9–10%, reflecting the near absence of new nuclear power plant commissioning. However, there are signs in 2026 that nuclear energy may see a resurgence in the 2030s with new power plant construction in Asia and the new generation of Small Modular Reactors.

The expectation in the middle of the third decade of the 21st century is that electricity demand will continue to grow at a robust rate, driven by the following factors:

1. The increasing global population, which is projected to top the 9.2 billion mark by 2040.
2. The increasing utilization of A/C systems for cooling, especially in the developing world. With 82% of the global population living in latitudes between 40° south and 40° north, with the global average temperature gradually rising, and with the growing economic prosperity in several developing nations, the A/C systems will continue to proliferate. A rather unfortunate fact in A/C proliferation is that, when older buildings in the developing world are refurbished to install A/C units, the owners' choices fall on the cheaper, small and inefficient individual units, as shown in the apartment complex of Figure 1.8. The smaller and less efficient units, which

FIGURE 1.8 A building in Mindelo, Cavo Verde Islands, refurbished with the smaller and rather inefficient A/C units.

are ubiquitous in several cities with older housing stock, consume substantially more electricity (up to three times more) than the efficient and larger central A/C units.

3. The continuing global proliferation of EVs. Even though the growth rates of EV sales have dropped in the USA and a few countries of the EU [15], the global growth of EV sales is strong, and EVs are expected to account for 40% of the global passenger-vehicle fleet by 2040 [16]. A silver lining here is that every EV in use reduces the corresponding demand for petroleum and the associated CO_2 emissions. If the EV charging times can be controlled (e.g., charge over a long time during the night) the charging processes can be more efficient, and most of the charging can be accomplished using the spare capacity of conventional power plants to generate electric power.
4. An increase in industrial production worldwide, including "green ammonia" for fertilizers; the production of metals such as aluminum and steel; and the manufacturing of appliances, furniture, and other implements. A growing and more affluent world population needs more food and more electric power to produce the food and the variety of consumer goods it desires.
5. The dawn of the AI era that demands electricity-voracious data centers in the developed nations. A preliminary estimate by the Electric Power Research Institute for the USA alone is that an additional 113 GW of electric power generation units (close to 10% of the current capacity) will be needed by 2030 for the development of adequate AI capability [17]. This may be a conservative estimate, since the business community's estimate is close to 400 GW, with at least 80 GW power being constructed in the period 2025–2030 [18, 19]. The surging data center investment has the potential to aggravate the national electricity supply. With more than 90% of data center operators citing power availability as their top concern for the growth of the AI industry, diverting power to their centers will adversely affect the average consumer. Making electricity available for data centers has also become a challenge for existing electricity grids: while new data center developments take 3–6 years to complete, new grid infrastructure typically takes between 5 and 15 years from planning to completion [20].

If the higher growth rates persist, the global electricity demand will almost double by 2040. Since electricity is a tertiary energy source generated by other primary sources, significant quantities of additional PESs must become available for the generation of the growing electricity demand. Renewable energy generation units (especially wind and solar) will continue their fast pace of growth but by themselves will not be able, in a short time, to generate all the additionally required electricity. As it will become apparent in Chapter 2, the lack of energy storage capacity is an impediment to the higher market penetration of the less dependable wind and solar energy. The contributions of hydraulic and geothermal energy, which do not need power storage, will be limited because the development of these two energy sources is not economically feasible in most regions of the planet.

There are signs (primarily in Asian countries) that the non-CO_2-emitting nuclear energy will also contribute to the additional electricity generation of the future.

Several large nuclear power plants (with power capacity in the order of 1,000 MW) are in the construction and commissioning stages in China, India, and South Korea. The smaller modular nuclear reactors (power generation capacity 20–200 MW) also appear to be a promising source of electric power for the future, especially in powering the large data centers.

Fossil fuels will continue to provide a high fraction of the global electricity generation. It is highly likely that the additional global electricity demand will be partly met by coal (which is available in most developing countries, including India and China) and by natural gas, with the contribution of the latter expected to significantly increase at least until 2040. However, this path implies higher quantities of CO_2 emitted in the atmosphere – an undesirable global sustainability outcome.

1.5 RESOURCES, RESERVES, AND RENEWABLE ENERGY

Fossil fuels and nuclear fuels – coal, natural gas, petroleum, uranium, thorium, etc. – were formed in finite quantities eons ago and are currently extracted from the earth's crust at rates significantly exceeding their replenishment rates. Fossil fuels and nuclear energy resources will be depleted at some point in the near or more distant future. Herein lies the *energy challenge* contemporary humans are facing: unless the human society develops sufficient alternative energy sources (preferably renewable) before the depletion of the current primary sources, we will experience severe energy shortages in the future that will jeopardize our civilization.

The crust of the planet holds significant quantities of fossil fuels, but not all of them can be economically mined. While developing an oil field that contains two hundred million barrels of oil is a worthwhile and profitable endeavor, it makes no sense to develop a deep mine for the extraction of only 15 tons of anthracite, worth less than $10,000. Economics and the cost of mining play important roles in the classification and possible utilization of energy resources. A small fraction of the energy resources is located in known regions, closer to the surface of the earth, and may be extracted economically with existing technology. This fraction is referred to as *proved* or *proven reserves.* Another part of resources is known to exist but may not be recovered economically at present. Such resources are classified as *potential reserves.* If the prices of fossil fuels rise or if a technological breakthrough is achieved, the potential reserves are re-classified as proven reserves and become available to be profitably extracted. The potential reserves are classified in two subcategories as *probable reserves* and *possible reserves,* with the probable reserves more likely to be recovered than the possible reserves.

In addition to the reserves there are quantities of energy resources, typically in regions close to the proven and the potential reserves, that have not been fully explored and quantified. From past experience with resource mining, it is reasonably assumed that, after scientific exploration and quantification, such resources may be discovered and recovered in the future. These are the *inferred resources.* Finally, there are additional resources that might exist in the crust of the earth, but they have not been discovered yet and may or may not be extracted economically. These resources may become available in the future, and they are classified as *undiscovered* or *hypothetical resources.*

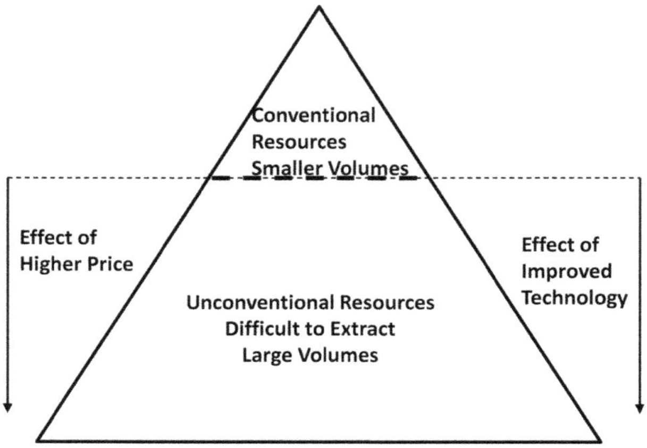

FIGURE 1.9 The triangle of energy resources and the effects of prices and technology.

It is apparent that the total resources by far exceed the reserves. Engineers have suggested [21] that the total quantity of an energy resource may be schematically depicted in a triangle as in Figure 1.9. The *conventional resources* represent the proven and a large part of the potential reserves, which are reasonably expected to be economically mined in the near future. The mass of the conventional resources is estimated at 10–20% of the total resources and is represented by the upper part of the triangle that has the smaller area. The bottom of the triangle with the larger area represents the quantity of the resources that exists in the crust of the earth, but may not be extracted at present, either because it is not economical to extract it or because the necessary technology has not been developed. These are the *unconventional resources,* which are estimated to be 5–10 times higher than the conventional resources. For petroleum, the unconventional resources include shale oil and viscous or heavy oil in low-permeability reservoirs.

Increased prices and improved technology shift the dividing line toward the bottom and increase the quantity of proven reserves. Until the early 1960s, the oilfields in the North Sea, between Scotland and Norway, were classified as potential or inferred resources [22]. With the technological advances in offshore exploration, drilling, and oil production and with the significant increase of crude oil prices during the mid-1970s, it became economical to extract these resources. Hence, the North Sea resources were promoted to proven reserves, and by 2000 they were producing petroleum. Natural gas resources have followed similar trends: the increased price of the natural gas in combination with horizontal drilling and hydraulic fracturing (fracking) triggered the economic extraction of gas from shale formations, which were not considered as reserves until the 1980s.

A word of caution on energy reserves and resources: There is a strong dependence of the reserves-resources classification on current technology and energy prices. In addition to the scientific uncertainty about their total quantity (mass or volume) several energy corporations and national governments have intentionally fudged the numbers in the past to underestimate or overestimate their energy reserves and resources for commercial and political purposes. For these reasons, reports on reserves and resources fluctuate significantly, and many reports of the past have proven to be unreliable [23, 24].

1.5.1 Sustainable Energy

A reliable supply of affordable energy is essential for the global society. Securing their future energy supply is one of the principal considerations for the governments of all nations. Energy forms have become commodities on which the economic growth of the nations is based. Energy sufficiency via domestic production and imports has played an important role in regional and global political and military conflicts since the beginning of the 20th century. The control of vital energy supplies – petroleum in particular – has been the cause of several wars [25, 26].

Because fossil fuels are expected to be exhausted in the future and, in addition, are the causes of global climate change, using alternative energy sources (which include all the renewables as well as nuclear energy) has become important for climate stabilization, the continued economic development of nations, and the sustenance of our civilization [11]. In a sustainable future renewable energy sources will replace a high fraction of fossil fuel consumption and will generate sufficient energy for the human society to maintain its standards of living. Factors that will affect and may limit the quantity and use of renewable energy are:

a. *Resource Availability* is a limitation for all energy forms. For example, if there are no geothermal resources within a region, electricity cannot be generated by geothermal energy. Similarly, if there are no high-slope rivers, hydroelectric energy is out of the question. The availability of material resources also affects inexhaustible energy resources: with the current state of technology, a scarcity of rare earth elements (in particular gallium, selenium, and indium) would hamper the manufacturing of photovoltaic (PV) cells that are essential for the conversion of the abundant solar energy to electricity.
b. *Technology Availability* has been a limitation (partial or complete) since the invention of the first locomotive. Harnessing fusion energy, where the limited fundamental knowledge blocks the development of fusion-generated electricity is an example of complete lack of technology availability. The manufacturing of PV cells from common materials; the manufacturing of efficient, reliable, and durable fuel cells; the development of gas turbines with maximum efficiencies; and the production of convenient, high-capacity, and economical energy storage systems are examples of partial technology availability.
c. *Economic/Financial Limitations* include the cost of energy production methods and the ability to finance projects that will produce energy. In a market-oriented society informed consumers will choose the least expensive alternatives to satisfy their energy needs. If renewable energy is to be widely utilized, its costs should be competitive with the cost of energy from other sources. This may happen either because new technology enables the development of more efficient wind and solar conversion systems or because the cost of fossil fuels significantly increases.

d. *Socio-Political Limitations* include the perceptions of the population about forms of energy and the political-regulatory framework that promotes or bans certain forms of energy. Halting the development of nuclear energy and shutting nuclear power plants in several countries is an example of such limitations. Renewable energy has enjoyed favorable public opinion – primarily because of its cleanliness and environmental benefits – and has been assisted by national policies that offer financial incentives for its development. A significant boost to renewable energy investment and production will be the adoption of binding international treaties to limit the global CO_2 emissions.

1.6 A NOTE ON UNITS

Energy has many faces – heat, mechanical work, electricity, nuclear, etc. – and academic disciplines on energy have been continuously developed in several countries since the Industrial Revolution. The result is a plethora of units that easily confuse the novice as well as the veteran: ergs, joules, Btus (or BTUs), kilowatt-hours, therms, calories (and Calories, with capital C, that represent 1,000 calories), cubic feet of gas, and barrels of oil are among the myriads of units used in the energy industry. The international system (*Systeme Internationale* or *SI*) of units has adopted the joule, J, [26] as the unit of energy. The joule is defined as the energy needed for a constant force of 1 newton (1 N) to move its point of application by 1 meter. In electrical applications 1 J is performed when a charge of 1 coulomb (1 Cb) moves through an electric potential difference of 1 volt (1 V). The joule is a rather small amount of energy with one kilowatt-hour, the typical electricity unit for homes, being equal to 3.6 million joules (1 kWh = 3.6×10^6 J). Single households use billions of joules annually and a nation's demand for energy is measured in tens of exajoules (EJ) with 1 EJ = 10^{18} J.

Since the advent of the Industrial Revolution a variety of other units have been defined and are still used in several applications. Among the more common units one encounters in the 21st Century are as follows [27]:

a. The British thermal unit, which is abbreviated as Btu (and less often as BTU) is defined as the quantity of heat input to increase the temperature of 1 lb of water from 59.5 to 60.5°F. 1 Btu is equal to 1,055 J and is often approximated as 1,000 J (1 kJ).
b. A larger energy unit in the British system is the therm (1 Therm), and it is equal to 100,000 Btu. Therefore, 1 Therm is equal to 1.055×10^8 J.
c. The calorie (cal) is an older unit that is still used in thermal and food applications and is equal to 4.184 J. 1 cal is the amount of heat required to raise the temperature of 1 gram of water from 14.5 to 15.5°C. When it comes to nutrition, one must be careful: The predominant unit is 1 Calorie (abbreviated as 1 Cal, with capital C). One Cal is equal to one thousand cal (1 Cal = 1,000 cal).
d. An extremely large amount of energy is the Quad (1 Q), a unit commonly used to measure the amount of energy consumed by regions and nations. 1 Q is equal to 10^{15} Btu, which is equal to 1.055×10^{18} J. Therefore, one Quad is approximately equal to one exajoule ($1 \text{ Q} \approx 1$ EJ).

A series of energy units, often used in the fossil fuel industry, is related to the chemical energy content of the different fossil fuels (coal, crude oil, and natural gas). Because the chemical composition and energy content of actual fuels – e.g., sweet East Texas crude, West Texas Intermediate (WTI), Saudi Arabian crude, etc. – largely depends on the chemical composition and the location of the fuel extraction, these units of energy have been defined according to international convention. Frequently used units of this kind and their SI equivalents are:

a. 1 ton of coal equivalent (1 tce) is equal to 2.931×10^{10} J.
b. 1 barrel of oil (1 bbl) is equal to 6.119×10^{9} J.
c. 1 ton of oil equivalent (1 toe) is equal to 10^{10} cal or 4.184×10^{10} J.
d. 1 cubic foot of natural gas (1 scf), measured at standard pressure and temperature (1 atm and 25°C) is equal to 1.072×10^{6} J. 1 scf is often approximated as 10^{6} J.

The unit of power (energy per unit time) in the SI is the watt, W, defined as 1 joule per second. When an engine consumes or produces one joule of work per second, its power is one watt (1 W). For larger engines the kilowatt (1kW = 1,000 W) and the megawatt (1 MW = 1,000,000 W) are used. Another commonly used power unit in the transportation industry is the horsepower, hp, which is equal to 745.7 W.

Multiples of all the SI units (including the joule, the watt, and the watt-hour) have also been defined by international convention. Prefixes that denote these multiples in the range 10^{-9} to 10^{18} are denoted as follows:

- 1 nano (n) denotes 10^{-9} (e.g., 1 nJ 10^{-9} J).
- 1 micro (μ) denotes 10^{-6} (e.g., 1 μJ = 10^{-6} J).
- 1 milli (m) denotes 10^{-3} (e.g., 1 mJ = 10^{-3} J).
- 1 kilo (k) denotes 10^{3} (e.g., 1 kJ = 10^{3} J).
- 1 mega (M) denotes 10^{6} (e.g., 1 MJ = 10^{6} J).
- 1 giga (G) denotes 10^{9} (e.g., 1 GJ = 10^{9} J).
- 1 tera (T) denotes 10^{12} (e.g., 1 TJ = 10^{12} J).
- 1 peta (P) denotes 10^{15} (e.g., 1 PJ = 10^{15} J).
- 1 exa (E) denotes 10^{18} (e.g., 1 EJ = 10^{18} J).

The prefixes are not specific to energy units and are applied to all other units of SI. Thus, 1 MW = 10^{6} W, and 1 TWh = 10^{12} Wh. It must be emphasized that the SI units and their multiples follow a precise notation, including the use of low case and capital letters. Errors in notation between lowercase letters and capital letters may result in calculation errors of several orders of magnitude: While mJ/mJ = 1 (by definition as an identity), MJ/mJ = 10^{9}. We all have to be very careful and precise with the notation of units! Table 1.1 shows the conversion factors for several energy units that are commonly used. The symbol E in the Table denotes the exponent of 10, as in most calculators. Following this table, 1 cal = 4.18 J, and 1 ton of coal is equal to 2.78×10^{7} Btu. The diagonal terms of the table are equal to one by definition, and each term is equal to the reciprocal of its diagonally mirror term.

To avoid "confusion by units" this book has been written using as few units as possible. Most Figures on energy demand and supply in this book are consistently in

TABLE 1.1
Conversion Factors for Energy Units

	Calorie	erg	J	Btu	Gallon of gasoline	Barrel of oil, bbl	Ton of oil, toe	kWh	ft-lb	Therm	Ton of TNT	Quad, Q	Ton of coal, tce	scf of nat. gas
calorie, cal	1	4.19E+07	4.184	3.97E-03	3.17E-08	6.84E-10	1.0E-10	1.16E-06	3.09	3.97E-08	1.00E-09	3.97E-18	1.43E-10	3.90E-06
erg	2.39E-08	1	1.00E-07	9.48E-11	7.58E-16	1.63E-17	2.39E-18	2.78E-14	7.37E-08	9.48E-16	2.39E-17	9.48E-26	3.41E-18	9.33E-14
joule, J	0.24	1.00E+07	1	9.48E-04	7.58E-09	1.63E-10	2.39E-11	2.78E-07	0.74	9.48E-09	2.39E-10	9.48E-19	3.41E-11	9.33E-07
Btu	252	1.06E+10	1,055	1	8.00E-06	1.72E-07	2.52E-08	2.93E-04	7.78E+02	1.00E-05	2.52E-07	1.00E-15	3.60E-08	9.84E-04
gallon of gasoline, gal	3.15E+07	1.32E+15	1.32E+08	1.25E+05	1	2.16E-02	3.15E-03	3.66E+01	9.73E+07	1.25	3.15E-02	1.25E-10	4.50E-03	123.1
barrel of oil, bbl	1.46E+09	6.12E+16	6.12E+09	5.80E+06	46.4	1	0.1462	1.70E+03	4.51E+09	5.80E+01	1.46	5.80E-09	2.09E-01	5,708
ton of oil, toe	1.0E+10	4.187E+17	4.187E+10	3.97E+07	317.8	6.84	1	1.16E+04	3.09E+10	3.97E+02	10	3.97E-08	1.43	3.91E+04
kWh	8.60E+05	3.6E+13	3.60E+06	3,412	2.73E-02	5.88E-04	8.62E-05	1	2.66E+06	3.41E-02	8.61E-04	3.41E-12	1.23E-04	3.358
Foot-pound ft-lb	0.32	1.36E+07	1.36	1.29E-03	1.03E-08	2.22E-10	3.24E-11	3.77E-07	1	1.29E-08	3.24E-10	1.29E-18	4.63E-11	1.27E-06
Therm	2.52E+07	1.06E+15	1.06E+08	1.00E+05	0.8	1.72E-02	2.52E-03	29.31	7.78E+07	1	2.52E-02	1.00E-10	3.60E-03	98.42
ton of TNT	1.00E+09	4.18E+16	4.18E+09	3.97E+06	31.73	0.68	0.1	1162	3.09E+09	39.66	1	3.97E-09	1.43E-01	3,899
Quad, Q	2.5E+17	1.06E+25	1.06E+18	1E+15	8.00E+09	1.72E+08	2.52E+07	2.93E+11	7.78E+17	1.00E+10	2.52E+08	1	3.60E+07	9.84E+11
ton of coal, tce	7.00E+09	2.93E+17	2.93E+10	2.78E+07	2.22E+02	4.79E+00	0.6993	8.14E+03	2.16E+10	2.78E+02	7.00	2.78E-08	1	2.73E+04
scf of nat. gas	2.56E+05	1.07E+13	1.07E+06	1,016	8.12E-03	1.75E-04	2.56E-05	0.2978	7.88E+05	1.02E-02	2.56E-04	1.02E-12	3.66E-05	1

toe or EJ, while the units for electricity are in the commonly used units, kWh, MWh, or TWh. The reader is able to make immediate comparisons among the figures, without the need for conversions.

NOTES

1 Feynman, R.P., and Sands, M.L., 1965, *The Feynman Lectures on Physics*, Addison-Wesley, Reading, MA.
2 The representative nations whose data are depicted in Figures 1.2, 1.3, and 1.4 are as follows: Argentina, Australia, Austria, Bangladesh, Bolivia, Brazil, Canada, China, Congo, Costa Rica, Cuba, Egypt, Ethiopia, France, Gabon, Germany, Greece, Haiti, India, Iran, Iraq, Japan, Kenya, Mexico, Russia, Senegal, South Africa, Spain, Sri Lanka, Turkey, UK, USA, and Zambia. The world average is also shown. To avoid graph cluttering, not all the nations are labeled.
3 Data from https://ourworldindata.org/grapher/energy-use-per-person-vs-gdp-per-capita
4 The OECD group of countries comprises most of the European Union countries, Switzerland, Canada, USA, Chile, Israel, Mexico, Turkey, South Korea, Australia, and New Zealand. Other countries with robust economies have joined the group periodically. The countries in this group are sometimes referred to as the "developed" countries.
5 United Nations Human Development Report 2015, United Nation Human Development Program, New York, 2015.
6 See https://hdr.undp.org/data-center/human-development-index#/indicies/HDI
7 See https://hdr.undp.org/data-center/documentation-and-downloads
8 Data from https://ourworldindata.org/energy-production-consumption and https://www.iea.org/data-and-statistics/data-product/world-energy-statistics-and-balances.
9 IEA, *Global Energy Review – 2025*, IEA publications, Paris, France.
10 Piementel D., 2003, Ethanol Fuels: Energy Balance, Economics, and Environmental Impacts are Negative, *Natural Resources Research*, 12, 127–134.
11 Michaelides E.E., *Alternative Energy Sources*, Springer, Berlin, 2012.
12 IEA, *Electricity 2024 – Analysis and Forecasts to 2026*, IEA publications, Paris, France.
13 It is a consequence of the 2nd Law of Thermodynamics that not all heat may be converted to mechanical work (electric energy is equivalent to mechanical work).
14 Smil V., *How the World Really Works*, Viking, 2022.
15 Mackrael K., and Wilmot, S., 2025, European Union Reverse Course on its Combustion-Engine Ban, *Wall Street Journal*, December 17.
16 Global Electric Vehicle Sales Set for Record-Breaking Year, Even as US Market Slows Sharply, *Bloomberg Reports*, June 18, 2025.
17 EPRI Study: Data Centers Could Consume up to 9% of U.S. Electricity Generation by 2030, May 29, 2024.
18 Hiller, J., 2025, AI's Bring Your Own Power, *Wall Street Journal*, October 15.
19 Leppert R., 2025, What we know about energy use at U.S. data centers amid the AI boom, *Report of Pew Inst.*, October 24.
20 IEA, *World Energy Investment 2025*, 10th Edition, IEA publications, Paris, France.
21 Masters, J.A., 1979, Deep Basin Gas Trap, Western Canada, *AAPG Bulletin*, 63(2), 152–181.
22 Foley, G., 1976, *The Energy Question*, Penguin Books, Middlesex, England.
23 Hubbert, M.K., 1971, Energy Sources of the Earth, *Scientific American*, 224, 31–40.
24 Laherrere, J.H., 1995, World Oil Reserves – Which Number to Believe? *OPEC Bulletin*, February 1995, 9–13.
25 Grathwohl, M., 1982, *World Energy Supply*, deGreuter, Berlin.
26 According to international convention, the SI units are spelled with a low case letter (kelvin, joule, watt, coulomb) and are abbreviated by capital letters (e.g., K, J, W, Cb).
27 Michaelides, E.E., *Energy, the Environment, and Sustainability*, CRC Press, 2018.

2 Environmental Effects of Energy Consumption

Having established the importance of reliable and affordable energy, including fossil fuels, for humans, this chapter examines the warming effects of atmospheric carbon dioxide. The concepts of weather and climate are differentiated; data on the average global temperature and the atmospheric concentration of CO_2 spanning more than twelve decades are presented; and the cause-effect relationship between the two is explained, using several sets of data and graphs. The analysis is extended to include the effects of historical CO_2 emissions by the industrialized countries as well as the contribution of a continuously increasing global population, primarily due to the developing world. Since the scientific community was successful in eliminating other environmental threats of the past – acid rain, atmospheric lead, and the ozone hole – the methods that contributed to the elimination of the threats are presented, and the uniqueness of the current Global Climate Change threat is contrasted with the past successes. Because of the global notoriety of the climate change subject and the attention it has received from journalists, activists, politicians, and self-proclaimed experts, this chapter concludes with an exposition of several myths that were promulgated from all sides about the certainty of the warming effect as well as some outrageously catastrophic predictions.

2.1 THE GREENHOUSE EFFECT

It is a consequence of the laws of Physics (and widely experienced by all) that when a system receives more heat (energy) than it rejects, the temperature of the system rises (or something within the system melts or evaporates). On the contrary, if the system rejects more heat than it receives, its temperature falls. The earth's biosphere continuously receives heat from the sun in the form of solar irradiance and, simultaneously, radiates in all directions toward the colder outer space. When the incoming solar energy exceeds the outgoing energy, the temperature of the biosphere increases. This is observed almost every day during the daytime, in particular at the solar noon when the irradiance is highest. During the nighttime, the earth continues to emit energy into the outer space, while the solar irradiance diminishes, and so the temperature of the biosphere drops. For the same reason, because the days are longer in the summer and more solar energy enters the local biosphere during the day, the average summer temperatures are higher than the average winter temperatures.

The planet's surface and atmosphere have an effect on the energy exchanges between earth and the outer space. Clouds and ice sheets reflect a great deal of the solar energy back to the outer space during the day. The same clouds reflect back to the earth a fraction of the terrestrial radiation during the night – the reason cloudy winter nights are, in general, warmer. Snow and ice reflect 92% of the solar

DOI: 10.1201/9781003674658-2

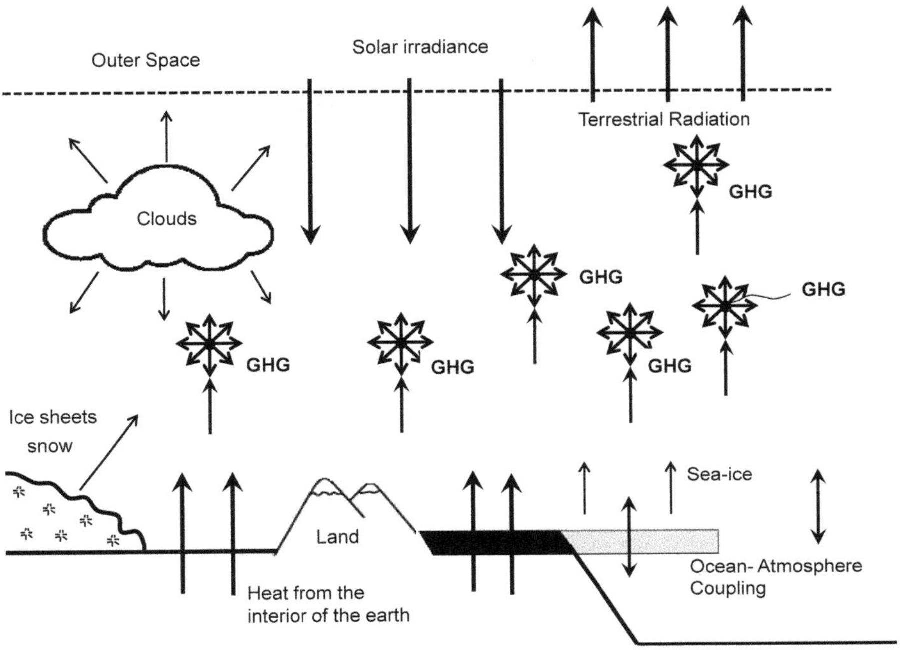

FIGURE 2.1 A schematic diagram of energy exchanges in the biosphere.

irradiance, while the built urban environment absorbs more than 80% of this irradiance. The surface waters (sea, lakes, and rivers) are like a heat reservoir that may absorb or give up a great deal of energy to the biosphere. Figure 2.1 is a schematic diagram that shows the energy exchanges between the biosphere, the sun, the outer space, and the earth's interior, which is at high temperature and conducts heat to the surface. Several of the land features, the water, and the atmosphere that affect the energy exchange between earth and space are also shown. Of particular interest in these energy exchanges are the molecules of the greenhouse gases (GHGs) that exist in small quantities in the atmosphere. The GHGs allow the solar irradiance to pass but selectively absorb part of the earth's radiation and reflect almost half of it back to the earth. The GHGs' mechanism is the same as that of a greenhouse cover that keeps the interior temperature higher than the outside temperature to increase vegetable growth and yield inside the greenhouse. By their selective emission-absorption properties, the GHGs contribute to a higher temperature in the biosphere.

It is of interest that, if the GHGs were entirely absent from the atmosphere and did not return to the biosphere some of the terrestrial radiation, the earth's temperature would have been significantly lower. Climate models conclusively show that, in the absence of GHGs, the average temperature of the biosphere would have been approximately 33°C (59°F) lower than its current average. At such temperatures, the oceans and the surface waters would have been permanently frozen, and the planet would have been inhospitable to life in its current forms. Essentially, the effect of the GHGs is to create a "blanket" that has warmed the biosphere for millennia and supported the lifeforms in the planet.

Carbon dioxide (CO_2), methane (CH_4), and water vapor are naturally occurring GHGs. They are also the ones that contribute most to the warming of the biosphere. While the low concentration of GHGs is necessary for the life to exist on earth, substantially higher concentrations of these gases and the addition of artificial GHGs further increase the average temperature of the biosphere. Effectively, the higher concentration of the GHGs thickens the "blanket" that made the biosphere hospitable to life forms and supported this life for eons. Environmentalists and ecologists warn that if the concentration of GHGs in the atmosphere is left unchecked and continues to increase, significant climatic effects will be triggered on the planet with possible disastrous consequences for the ecology and the human economic activities. It is rather ironic that the very gases, which have kept the temperature of our planet at levels that support and nourish life, now threaten the ecosystem and the global human society.

The relationship between the GHGs (in particular CO_2 concentration in the atmosphere) and global temperature is not a new and recently developed scientific theory. A few years after the beginning of the Industrial Revolution, in 1824, Joseph Fourier – a French savant who is widely regarded as the "father" of the *Heat Transfer* scientific discipline – observed that certain atmospheric gases absorb and reflect back a fraction of the earth's infrared radiation [1]. John Tyndal, a British scientist of the 19th century who studied the subject of infrared radiation absorption in the atmosphere, wrote that variations in the atmospheric concentrations of water vapor and "carbonic acid gas" (CO_2) will cause climate change [2]. The Swedish chemist Svante Arrhenius, who was awarded the Nobel Prize in Chemistry in 1903, actually calculated the forcing effect of CO_2 in the atmosphere to be +4°C if the atmospheric concentration of this gas doubled to 560 parts per million (ppm) and −4°C if the concentration halved (to 140 ppm). He also concluded that the warming effect would be higher in the polar regions than in the tropics and that the warming of the nights will be higher than that of the days [3, 4]. Contemporary calculations [5] on CO_2 and climate change with supercomputers show that the doubling of CO_2 atmospheric concentration (to 560 ppm) will increase the average global temperature between 2.6 and 3.8°C, and that the heating effects will be more pronounced in the polar regions, as Arrhenius predicted in the early 20th century.

2.1.1 Fossil Fuels and Carbon Dioxide Emissions

The fossil fuels – coal, petroleum, and natural gas – were formed in the earth's interior over millions of years from plants and other organisms during the Carboniferous Period, roughly 300 million years ago. All fossil fuels are composed of carbon and other atoms, chiefly hydrogen. When the fossil fuels burn to generate heat, the carbon atoms combine with atmospheric oxygen to produce carbon dioxide (CO_2) that is typically discharged in the atmosphere. The amount of carbon dioxide produced to deliver a quantity of heat varies with the type of molecules of the fossil fuels. Carbon, the main constituent of coal, produces 0.11 kg of CO_2 to deliver 1 MJ of heat; octane, one of the main constituents of gasoline, produces 0.064 kg of CO_2

to deliver the same 1 MJ of heat; and methane, the main constituent of natural gas, produces 0.05 kg of CO_2 to deliver 1 MJ of heat (for those who prefer the British units 1 MJ is approximately equal to 1000 Btu). The various types of coal (anthracite, bituminous, lignite, etc.) produce more CO_2 than other fossil fuels for the delivery of a given quantity of heat.

For several millennia, the principal fuel of humans was wood and other forms of plant biomass that do not burden the atmosphere with additional quantities of CO_2. Plants absorb CO_2 molecules from the atmosphere to form wood and when this wood is burned (e.g., in stoves or fireplaces) all the released CO_2 simply returns to the atmosphere, from where it originated. During the several millennia, when wood was the principal source of heat for humans, there was an equilibrium between the amounts of CO_2 absorbed and released, so that the average concentration of this gas in the atmosphere remained almost steady, hovering around 280 ppm.

This equilibrium was disturbed by the onset of the Industrial Revolution, the invention and proliferation of heat engines, and the need to burn fossil fuels that were stored for millions of years inside the earth. From that point onward, fuels with carbon content were mined, processed, and burned to produce and discharge additional quantities of CO_2 in the atmosphere. Fossil fuel combustion continuously increased with the result that in 2025 the CO_2 emissions in the atmosphere topped 39 billion tons. Figure 2.2 shows the dramatic increase of the global CO_2 emissions from 1900 to 2025 [6], together with the geopolitical events that affected the global energy consumption. It is apparent in this figure that the CO_2 emission growth significantly increased after 1950, when the automobile became a ubiquitous feature first in the USA, then in the other Organization for Economic Development and Cooperation (OECD) countries and, more recently, in the entire world. One can observe the similarities of the trends in Figures 1.5 and 2.2. During the period 1900–2025, the global primary energy consumption increased by a factor of 6.5, while the globally emitted CO_2 increased by a factor of 6.3. Obviously, there is a strong correlation between the two. One can also see that the dips in the two curves correspond to the same years,

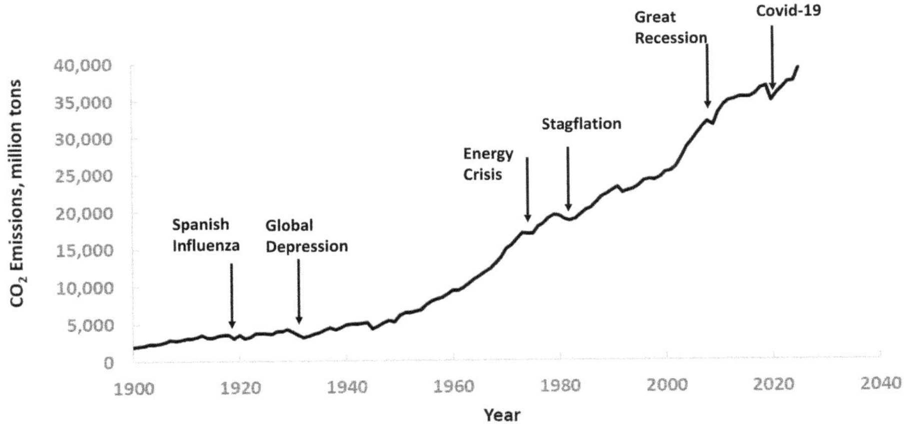

FIGURE 2.2 Global atmospheric emissions of CO_2 from 1900 to 2025.

clearly showing that the global consumption of energy also drives the global CO_2 emissions.

2.1.2 Other Greenhouse Gases

Carbon dioxide is not the only GHG. Water vapor (humidity) and methane are two of the other naturally occurring GHGs, while several man-made materials contribute to the greenhouse effect. Among the latter the most potent gases are: the nitrogen oxides that are formed in small quantities during fuel combustion in diesel and gasoline engines; and the various refrigerants currently used in refrigerators and air-conditioning systems, including the freons 11 and 12 (which have been banned in most countries) and the more modern hydrofluorocarbon refrigerants (such as the R-114, R-134a, and 142b), which are banned by the Kigali agreement (see Section 2.4.4). The atmospheric concentration of the other GHGs has also increased significantly since the 1960s due to human activities. The concentration of methane increased from 715 ppb (parts per billion) at the pre-industrial age to more than 2,000 ppb in 2025. Because of the internal combustion engines that power most vehicles, the atmospheric nitrogen oxide concentration increased from 270 to 319 ppb.

It must be noted that some of the other GHGs are significantly more potent than CO_2 in reflecting back to earth the infrared radiation [7]. For example, methane is 58 times more potent than CO_2 (per unit mass); and the refrigerant R-134a, which is the working fluid of most air-conditioning systems in 2025, is 4,530 times more potent than CO_2. However, the total emissions and the atmospheric concentrations of these gases are much lower than that of CO_2. The overall warming effect of all these gases combined is less than 20%, significantly lower than that of CO_2 [7]. This is the reason why most international agencies including the United Nations Intergovernmental Panel on Climate Change (IPCC) and the International Energy Administration (IEA) have primarily focused on the CO_2 emissions and continuously advise the governments of all nations to reduce their CO_2 emissions [8–9].

2.2 ATMOSPHERIC CO_2 CONCENTRATION AND GLOBAL CLIMATE CHANGE

The combustion of fossil fuels and the ever-growing global CO_2 emissions have significantly increased the concentration of this gas in the atmosphere. Two and a half centuries of fossil fuel burning (since the dawn of the Industrial Revolution) has resulted in a continuous increase of the atmospheric CO_2. This is shown in Figure 2.3 which depicts the evolution of the average CO_2 atmospheric concentration from 1750 – the beginning of the Industrial Revolution – to 2025 [7, 10]. For several millennia, before the Industrial Revolution, the concentration of this gas remained almost constant at about 280 ppm. The CO_2 concentration started gradually rising after 1750, reached 300 ppm in 1920, 400 ppm in 2015, and accelerated to 429 ppm in 2026, 53% higher than its historical, pre-industrial levels. Since CO_2 is one of the GHGs, its increasing atmospheric concentration effectively thickens the gaseous "blanket" that surrounds the earth and contributes to the temperature increase of the planet, a phenomenon that leads to climate change.

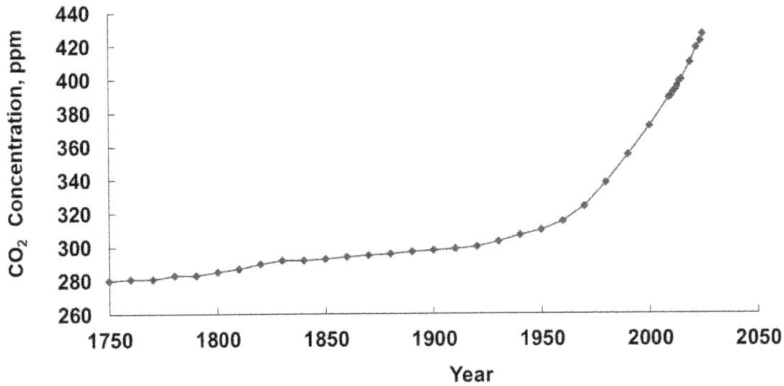

FIGURE 2.3 Average global CO_2 atmospheric concentration.

2.2.1 Weather and Climate

Before any discussions of the Global Climate Change (GCC), it is sensible to define our understanding of "climate" and distinguish between weather and climate. The *weather* is the short-term product of all the complex interactions between the sun and the earth's biosphere. Rain at 9:00 am on March 8, high temperature at 2:00 pm on July 17, and freezing temperature on February 5 are all weather phenomena. The weather is typically described by measurable variables such as the local temperature, pressure, rainfall, humidity, wind velocity, etc. It is a short-term phenomenon that results from the temporary thermal interactions between the solar radiation, the atmosphere and the hydrosphere [11]. The weather is predictable over short periods, e.g., a few days, but is unpredictable over long periods of time (no one can accurately predict what the temperature will be and if it will rain at 5:00 pm seven months from today).

The *climate* is the average weather with the average taken over a long time, typically a period of a few decades. Unlike the weather, climate is predictable in the long run: we know with certainty that the summer, ten years from now, will be warmer than the winter and the spring of the same year; that the average wind velocity in the spring will be higher than that of the summer; and that, in the northern hemisphere, there will be lower temperatures and more snowfall in January than in July.

While the weather is variable, the climate is stable, and most human economic activities depend on a stable and predictable climate. All agricultural activities that supply our food have been based for millennia on the predictability of the climate and the seasons. Food-producing plants remain dormant in the winters and grow in the springs and summers to be harvested. One may rely on the fact that the January days in Hanover will be colder than in the Canary Islands; the month of March in London will carry a significant amount of rain; that July in Fort Worth, Texas, will be hot and dry; and that the month of January will be warmer in Buenos Aires than in New York. Weather may bring rain on a particular October day in the Sahara Desert, but one can rely on the climatic fact that, during a given year or a given season, the rainfall in the Sahara will be substantially lower than the rainfall in the Amazon basin.

Weather changes frequently, but the weather phenomena are brief and do not significantly impact the ecology, the environment, and most human activities. On the contrary, climate changes, whether regional or global, significantly affect the environment, the ecosystems, and the human activities. For example, a long-term 5–7°C temperature increase in Nebraska and Kansas accompanied by drought will convert that region into a desert, will destroy the local economy, and will deprive the USA of its breadbasket. Permanently higher temperatures that cause polar ice melting and a 1 ft (0.3 m) rise of the sea level will have catastrophic consequences for Bangladesh, Louisiana, and the Maldives. The increase of the average temperature of the earth is a *de facto* GCC that has the potential to severely affect all human activities and the entire global economy.

2.2.2 Global Temperature Rise

The earth is a complex, highly nonlinear, dynamic system, where small changes in the atmospheric composition will cause substantial local and global weather deviations. When these deviations become permanent, the climate is altered followed by environmental and ecological effects. One of these global deviations has become apparent and has been confirmed by the scientists in the last thirty years: the average annual temperature of the biosphere shows a clear and increasing long-term trend in the last several decades. This is shown in Figure 2.4, which depicts the temperature deviation of the lower atmosphere [12]. The temperature deviation is defined with respect to the time-averaged temperature of the 20th century (that is, the zero in the graph corresponds to the average global temperature in the years 1901–2000). The dots in Figure 2.4 correspond to the temperature of the biosphere during the given year and have to do a lot with the weather during that year (for millennia some years have been hotter than others). The solid line corresponds to the ten-year averages of

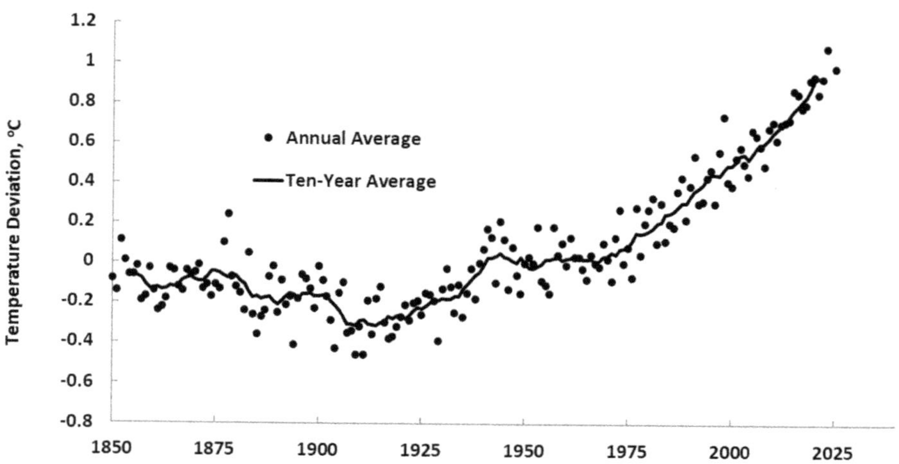

FIGURE 2.4 Average global temperature since 1850. The zero in the figure corresponds to the average 20th-century temperature.

the annual temperatures and is an indication of climate trends. A glance at Figures 2.3 and 2.4 proves that there is an obvious correlation between the atmospheric CO_2 concentration and the ten-year average global temperature deviation since the 1960s. This reinforces the cause-effect relationship between atmospheric CO_2 concentration and climate, which has been the subject of warnings from scientists since the times of Fourier [1] and Arrhenius [3, 4].

One also observes in Figure 2.4 that there is a noteworthy variability in the annually averaged data and a lesser variability in the ten-year averaged data. Because of combinations of weather phenomena, we have colder and hotter years (solar activity and planetary weather phenomena, such as the "El Niño" or "La Niña" sea currents explain the yearly weather variability). However, the ten-year averages show a lesser degree of variability and are more indicative of long-term trends associated with a changing climate. It is also observed in Figure 2.4 that there have been several periods ranging from a few years to twenty years (e.g., 1943 to 1963) when it appears that the temperature is almost constant or even slightly negative. This, in combination with the measurement uncertainty of the data, has led a few individuals – primarily non-experts in statistics and climatology – to doubt that the temperature of the biosphere is increasing in the long term. In order to draw conclusions on the climatic effects, one must be reminded that climate is the long-term average weather. The temperature data since the 1970s unequivocally indicate a significant upward trend, the signal of the GCC.

2.2.3 Potential Effects of Global Temperature Rise

Climatological models agree that the increasing concentration of the GHG's in the atmosphere will continue to have a warming effect on the earth's climate. While the current temperature rise is based on scientific measurements with low uncertainty, the magnitude of this rise in the future, and the other effects this will have on humans and the ecosystems are laden with substantial uncertainty. From the beginning, we must make clear that all future predictions are based on computer climate models. The parameters used in the models are inherently laden with some degree of uncertainty; and this uncertainty always influences the predicted future state of the planet.

It is rather unfortunate that model uncertainty is omitted in the way climatic effects are reported in the communications media, and the general public is daily confronted with sensational scenarios of a bleak future for the humanity, that can only be avoided if we immediately act with urgency. For example, a 2014 study, which received a great deal of attention in the press, urged immediate action because the 2°C level would be surpassed by 2036, even if the atmospheric CO_2 concentration were to only increase to 405 ppm [13]. However, in early 2026, the CO_2 concentration is 429 ppm, and the temperature increase is well under 2°C. Another doomsday study, which predicted that the global economic output would drop by 62% (we will all be dirty poor if this happened!) by the end of the 21st century and was published in a prestigious journal, was retracted one year later, because of faulty methodology and high uncertainty [14].

The uncertainty and dire predictions notwithstanding, it is noteworthy that the vast majority of climate models agree that, if the current trends in CO_2 emissions continue and the concentration of this gas exceeds the 450-ppm level, there will be a global temperature increase, close to 2°C, and maybe higher [9, 15]. The same models also predict regional and global climate impacts that have the potential to disrupt the human economic activities and need to be addressed by adaptation measures. The most important consequences of GCC, which are detrimental to the environment, the ecosystems, and the human economic activities are [9, 16]:

a. *Melting of the Polar Ice:* The expected temperature rise in the polar regions is higher than in the equatorial and temporal regions and this will cause the partial melting of ice caps. Since ice reflects 92% of the incident sunlight and the polar ice caps reflect a high fraction of the incident solar radiation, ice reduction in the polar regions will feed back in the global energy exchange and will further increase the temperature rise of the planet.
b. *Sea Level Rise:* The melting of the ice caps creates enormous masses of water that drains into the sea. In addition, as the oceans' temperature rises, thermal expansion increases the volume of the water in the oceans. The two effects cause a sea-level rise. It is difficult to accurately calculate the climate-induced sea level rise from actual coastal observations, because land subsidence, which has nothing to do with GCC, also contributed to the observed 0.15–0.20 m (6–8 inches) global sea level rise between the years 1950 and 2020 [16]. Similarly, the projected 0.25–0.30 (10–12 inches) future average sea-level rise in the USA by 2050 is subject to substantial uncertainty that leads to conflicting predictions. While the National Oceanic and Atmospheric Administration (NOAA) claims that there will be a 0.3 m sea level rise in Manhattan by 2050 [17], a report by the US Department of Energy (US-DOE) casts doubts about this and similar predictions [18]. Even with the use of the most powerful computers, the words of Homer still ring true: *no one ever knew the future*! In this case, we do not know the future effects of increasing CO_2 concentration with any degree of certainty.

   A high sea-level rise is arguably the most catastrophic potential effect of GCC. While the NOAA and US-DOE predict a modest sea-level rise, a recent European model predicts up to a 5-m rise (by 2150) if the CO_2 emissions are not substantially reduced and the West Antarctic ice sheet melts [19]. A lesser but significant [20] (more than 1 m) sea level rise will bring underwater large parts of the continents and will threaten coastal areas with frequent floods. Under the high seal level rise scenarios, not only cities such as Venice and New Orleans will be essentially underwater, but other large metropolises and thriving economic hubs, such as New York, Shanghai, Los Angeles, Karachi, Rio de Janeiro, and London, will be severely threatened. Many of the low-lying Indian and Pacific Ocean islands (e.g., the Maldives, Kiribati, and Tuvalu) will be inundated by storms; large coastal parts of the planet will become uninhabitable; and a large fraction of the earth's population will need to be relocated. Having personally (my family and I lived in New Orleans until 2005) experienced the chaotic conditions in the aftermath of hurricane Katrina, I fear that the permanent and uncontrolled displacement of millions of people worldwide is likely to break the societal bonds that form the nations and replace them with anarchy and destruction. Many countries are not equipped to respond to such dramatic events and may crumble under the chaos of displaced populations.

c. *Higher Intensity of Weather Phenomena:* Weather phenomena are primarily driven by temperature differences and, since these differences increase in magnitude, the intensity of weather phenomena will also increase. Tropical storms, cyclones, and hurricanes are expected to be stronger. Winter storms and polar vortices are expected to become more intense and harsher.

   It must be noted that there is not yet an independent scientific confirmation of this effect. The higher intensity of storms was the conclusion of analytical studies starting in the 1990s [21], but the high uncertainty associated with weather phenomena predictions makes almost impossible to confirm this effect in any past or current weather phenomenon. This is rather unfortunate, because several officials and some of the communications media invariably blame GCC for weather-associated calamities – from hurricanes in Florida to fires in California stoked by high winds. A moment's reflection proves that hurricanes and fires are weather phenomena that have afflicted these regions for millennia. Actually, the word "hurricane" originates in an ancient language of the Caribbean, while the Santa Ana winds have blown for ages in California. Whether or not the observed 1.1°C rise of the average global temperature in Figure 2.3 has significantly increased the intensity of these phenomena is still a matter of scientific debate [19].

d. *Regional Climate Change:* The average global temperature rise will be accompanied by regional temperature rises, some more significant than others. As a consequence, the climate of several regions will change. One must note, however, that the climate models, which attempt to predict these regional changes, are not sufficiently validated to consider such future predictions accurate. For this reason, the current regional predictions of the models are subject to high variability and uncertainty. Two predictions that have received a great deal of attention are: (a) the desertification of parts of the Midwest and the Great Plains in the USA, because of chronic drought; and (b) more frequent rainfall in the Sahara, which will transform the desert to fertile land. Such regional predictions of the models are uncertain and may not materialize in the future. However, scientific reasoning, common sense, and all the scientific models agree that the global average temperature will further increase, and regional climates may change. This will have unwelcome consequences on the current economic activities (especially agriculture) that depend on a constant climate, predictable seasons, and relatively predictable rainfall.

e. *Ocean Acidification:* Oceans absorb 25%–30% of the anthropogenic CO_2 emissions. Oceans and plants together absorb close to 50% of the emissions [22]. The continuous uptake of this gas that forms the weak carbonic acid increases the acidity of the oceans. However, one must bear in mind that the ocean water is alkaline, with its pH close to 8. The absorption of CO_2 does not "acidify" the ocean but makes the water less alkaline and closer to the neutral pH value of 7 [23]. While several publications advocate that the CO_2

uptake and the pH reduction will negatively affect the aquatic ecosystems [24], others contend that the ocean pH reduction has a negligible direct impact on fish and coral life behavior [25].

f. *Local Precipitation Changes*: Since local temperatures affect the atmospheric humidity and rainfall, there will be changes in the annual precipitation that will affect local water availability, agriculture, and animal husbandry.

g. *Positive Effects of Global Warming:* Not all GCC effects are negative. The growing CO_2 concentration in the atmosphere enhances the photosynthesis process, promotes plant growth and improves water use efficiency by plants. This has been called the "global greening" phenomenon [19] and has the potential to improve the agricultural yields as well as to mitigate the warming effects of GCC. Mitigation ensues because the increased plant evapotranspiration processes and the changed local air circulation have a decreasing effect on the local temperatures. During the period 1987–2017 the global leaf coverage increased by 8%, very likely as a result of the higher atmospheric CO_2 concentration [26].

When one examines the scientific measurements and facts about GCC and its consequences, it becomes apparent that there is a great deal of uncertainty about the expected effects of GCC. The earth is a highly complex system, and the climate of the future depends on a multitude of parameters and competing cause-effect chains that make accurate investigations and predictions extremely challenging. While the mass communications media prefer to present the conclusions of climate studies in a definitive and simplistic way to the general public, the IPCCs have always cautioned about the scientific challenges and the uncertainty surrounding all climate models. Cramer and his coworkers, who participated in the 2014 IPCC report, remarked:

> Two broad challenges to the detection and attribution of climate change impacts relate to observations and process understanding. On the observational side, high-quality, long-term data relating to natural and human systems and the multiple factors affecting them are rare. In addition, the detection and attribution of climate change impacts requires an understanding of the processes by which climate change, in conjunction with other factors, may affect the system in question [27].

2.3 WHO IS RESPONSIBLE FOR THE CO_2 ACCUMULATION?

Every year, and toward the end of the year, the *Conference of Parties* (COP) takes place, always in a different venue. The *Parties* in this conference are the signatory countries of the United Nations Framework Convention for Climate Change (UNFCCC) which first convened in New York, in 1992. The main objective of the UNFCCC and of the Conference is to stabilize the CO_2 concentration, encourage, and coordinate national efforts to combat climate change. Since the 2009 conference (in Copenhagen), a large number of the COP delegates focused on funding sustainability efforts in developing countries. The developed nations committed to establish a climate fund with the goal of raising 100 billion US$ annually by 2020 to combat

GCC and promote sustainability projects in developing nations. In 2024, this goal was raised to 300 billion US$ by the year 2035 [28].

Despite the fact that the national contributions to the climate fund did not timely reach the goal of $100 billion, the 2022 COP delegates (in Egypt) introduced claims for *loss and damage* – compensation of communities and persons for loss or damage to their property, due to intense weather phenomena attributed to GCC. Most important, developing nations that may be significant CO_2 emitters are exempted from contributing to this fund. This is effectively a mechanism for the transfer of wealth from the developed to the developing nations, regardless of their current and future contributions to the global CO_2 emissions. Because the developed nations are constantly blamed for the GCC effects it is worth examining the sources of atmospheric CO_2 emissions and, most importantly, if the climate fund will have any effect in reducing the global CO_2 emissions.

2.3.1 CO_2 Emissions – Total, Per Capita, and Cumulative

Since close to 80% of the energy consumed globally is generated from fossil fuels, it is self-evident that countries that use a lot of energy (as in Figure 1.2) also emit a lot of CO_2. Table 2.1 lists the twelve highest CO_2 emitting nations (2023 data [29]). Their emissions in 2000 are also included in the table as well as the percentage increase (or decrease if negative) in the period 2000–2023. Since global warming and the GCC are primarily 21st-century concerns, the last column indicates the result of the national efforts to combat the atmospheric CO_2 accumulation and GCC.

It is apparent that Table 2.1 primarily includes the most industrialized nations that started using fossil fuels earlier (e.g., USA, EU, Japan, Canada, etc.) and, also the most populous nations (e.g., P.R. China, India, USA, EU, Indonesia, etc.). Several of

TABLE 2.1
Highest CO_2 Emitting Nations in 2023 and Comparison with Emissions of 2000

Country	% of Total	Total 2023, tons	Total 2000, tons	% Increase
P.R. China	34.0%	13,259.64	3,666.95	+262%
USA	12.0%	4,682.04	5,928.97	−21%
India	7.6%	2,955.18	995.65	+197%
European Union	6.4%	2,512.07	3,563.26	−3%
Russia	5.3%	2,069.50	1,681.14	+23%
Japan	2.4%	944.76	1,248.81	−24%
Iran	2.0%	778.8	353.93	+12%
Indonesia	1.7%	674.54	299.09	+126%
Saudi Arabia	1.6%	622.91	265.24	+135%
Germany	1.5%	582.95	871.74	−33%
Canada	1.5%	575.01	543.04	+6%
South Korea	1.5%	573.54	474.16	+21%

the latter are still developing nations. The energy demand of the populous developing nations – and, consequently, their CO_2 emissions – are expected to increase in the near future. Of the industrialized countries in Table 2.1, all but Canada have reduced their CO_2 emissions in the period 2000–2023 (South Korea joined the industrialized nations group after 2000).

Another statistic that may help assign accountability for the GCC is the national emissions per person (*per capita*). Several nations emit high quantities of CO_2 because they have high populations. Table 2.2 shows the *per capita* emissions of several developed and developing countries.

One can easily observe in Table 2.2 that there is a large disparity of the CO_2 emissions among people of different nations. While the average American is responsible to 13.7 tons of CO_2 emitted annually, the average Mexican only contributes 3.75 tons. Based on such numbers, it would appear that Americans, Canadians, Australians and Saudis bear higher responsibility for the GCC than other people. It is also noted in Table 2.2 that several highly industrialized nations, such as the UK, France, Belgium, the Netherlands, etc., which generate a large fraction of their electricity from the (non-CO_2 producing) nuclear power plants score fairly low in the *per capita* emissions.

Countries that have industrialized earlier (Australia, France, Germany, Japan, Netherlands, UK, USA, etc.) have historically emitted a great deal of the atmospheric CO_2. The CO_2 molecules persist in the atmosphere for centuries, and this means that the emissions of this gas from the early days of the Industrial Revolution continue to contribute to the heating of the planet. A more reasonable way to determine CO_2 accountability is to compute the cumulative amount of CO_2 emitted by each nation, using historical emissions data. A study that compiled such data from 1850 to 2023 attributed 17.2% of the atmospheric CO_2 accumulation to the USA, 12.9% to the P.R. China, 10.3% to the EU countries, 6.8% to Russia, and 4.8% to India [30]. It is of interest that two developing but highly populous nations (India and P.R. China) have made it in this list – ahead of the industrialized Japan, France, and the UK – primarily because of their high recent emission history and their large populations. Another

TABLE 2.2
***Per Capita* Emissions of Several Countries (in Tons of CO_2)**

Country	CO_2 Emissions *per Capita*, tons	Population, Millions	Country	CO_2 Emissions *per Capita*, tons	Population, Millions
Australia	14.02	27	Iran	8.6	92
Bangladesh	0.73	176	Italy	5.18	59
Brazil	2.27	213	Japan	7.59	123
Canada	14.35	42	Mexico	3.75	132
Egypt	2.18	118	P.R. China	9.4	1416
France	4.14	69	Saudi Arabia	18.48	36
Germany	6.95	84	South Africa	6.29	65
Greece	4.97	10	Spain	4.49	48
India	2.05	1460	UK	4.41	69
Indonesia	2.4	286	USA	13.7	347

interesting fact is that when one ranks countries based on their cumulative emissions *per capita,* the green but sparsely populated islands of New Zealand come on top, followed by Canada, Australia, the USA, and Argentina.

2.3.2 CO_2 Emissions and World Population

It is not only the atmospheric concentration of CO_2 that significantly increased since the 1950s. The world's population exploded during this time period and the two are connected: more people living on the planet use more energy, which is generated by burning more fossil fuels and producing more CO_2. The rising living standards and increasing affluence exacerbate this effect (as shown in Figure 1.2) because not only the population numbers skyrocket but also do the energy and CO_2 emissions *per capita.* Figure 2.5 shows the global population (values on the right axis) together with the global CO_2 emissions (values on the left axis) in the period 1900–2025. During this period the global population rose from 1.65 to 8.2 billion, while the global CO_2 emissions skyrocketed! The trends in the two sets of data are very similar with a notable acceleration in both the population and the CO_2 emissions since the 1950s.

Simply put, a higher and more affluent global population emits more CO_2 and contributes to global warming.

Based on the high correlation between global population and CO_2 emissions, it is of interest to examine how developed and developing nations contributed to the population growth. Figure 2.6 depicts these contributions by showing the population of the countries that were members of OECD in 1975 (they are chiefly considered as the industrialized, developed nations) and the rest of the world's population [31]. The reason for choosing the OECD-1975 members is that the member nations of this Organization have increased and several of the new OECD members (e.g., Mexico, Columbia, Costa Rica, and others) are still considered as developing nations in the annual COP meetings.

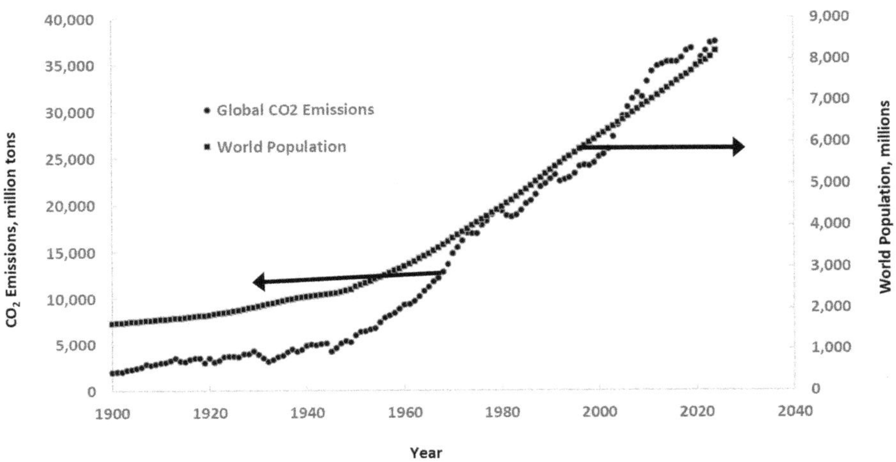

FIGURE 2.5 World population (right axis) and global CO_2 emissions (left axis) in the period 1900–2025.

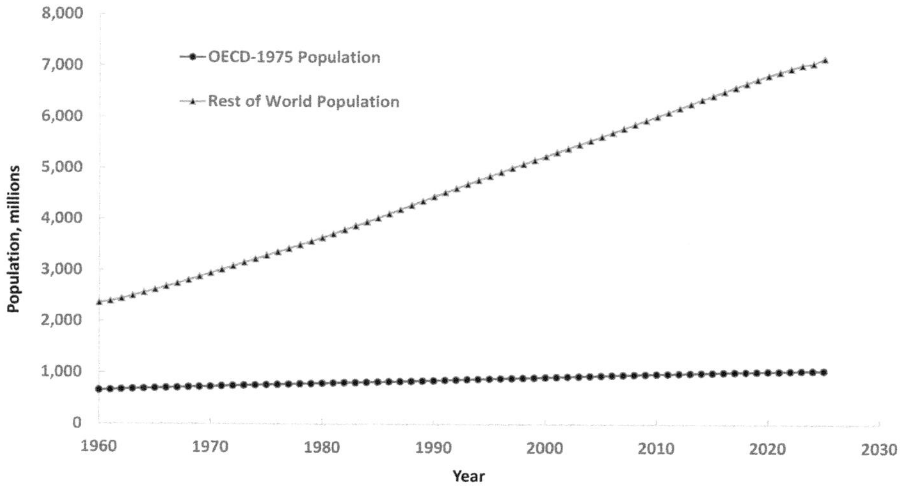

FIGURE 2.6 Populations of the OECD-1975 countries and of the rest of the world since 1960.

A glance at Figure 2.6 proves that, since 1960 when the warming trend of the planet accelerated, the population in the developed countries (1975 OECD members) increased by less than 60% (and most of the recent increase is due to emigration) but the population of the rest of the world has more than tripled. This places a considerable burden on the developing countries for accelerating the GCC and its potential threats. The earth is a planet with finite area and natural resources. A skyrocketing earthlings' population is placing substantial constraints on the resources of the planet, including the air we breathe and its CO_2 concentration.

Tackling the high *per capita* emissions of the developed nations without stopping the overpopulation of the planet will make the GCC problem worse. A simple exercise will prove it: let us assume that the two energy-voracious countries USA (population 340 million) and Canada (population 41 million) decide to bring their *per capita* CO_2 emissions from the current levels in Table 2.2 (13.70 and 14.35 tons, respectively) to 7 tons (approximately half the current emissions). At the same time India (population 1,451 million) and Indonesia (population 283.0 million) achieve higher prosperity and industrialization and their emissions increase from the current levels (2.05 and 2.4 tons, respectively) to 5 tons *per capita*. Simple arithmetic proves that the result will be a 2,437 million tons increase of the CO_2 emissions by the group of four countries, corresponding to an increase of 7% of the global emissions [32].

One cannot overlook the observation in Figure 2.5 that global population and accelerating global CO_2 emissions go hand-in-hand. Unless we tackle the problem of world overpopulation in the near future, we cannot solve the problem of the impending GCC effects. Even if the developed nations significantly curb their *per capita* emissions, the growing population in the developing world will thwart that concession.

The extensive contribution of world's population to the global warming raises significant doubts that the developed nations are the only culprits for GCC and its effects on communities and nations. The rapid growth of the population of several developing nations has caused internal migration and the crowding of cities. Several cities in the developing world that claimed less than a million inhabitants in the 1960s now have in excess of 10 million people. Overpopulation in these cities places excessive pressure on their environment, natural resources, and natural defenses against storms. In 2025, there were 17 metropolitan areas in the world with populations more than 20 million and another 34 with more than 8 million inhabitants [33]. The majority of these cities are in developing countries in coastal areas that make their populations vulnerable to weather phenomena. The overpopulation of coastal and more vulnerable regions raises questions that lead to extensive doubts about claims for loss and damage in crowded cities, due to weather events. One may succinctly express this issue with the question: Are the developed countries responsible (and, therefore, should pay) for a monsoon's loss and damage in Dhaka, Bangladesh, a city that grew from 336 thousand inhabitants in 1951 to more than 23 million by 2025?

2.3.3 The Failure of the Kyoto Protocol

The Kyoto protocol refers to an international treaty that was based on the principles of UNFCCC. It was adopted in 1997 in Kyoto, Japan, with the goal to reduce the global GHG emissions, primarily CO_2 emissions. The signatory nations established reduction targets for the industrialized nations, aiming at a 5% reduction below the 1990 levels (approximately 21,600 million tons of CO_2) to be achieved between 2008 and 2020. The protocol was based on the premise that industrialized nations were primarily responsible for the cumulative GHG emissions, placed higher responsibility on developed nations for the CO_2 reductions, and essentially exempted the developing nations from any restrictions on their emissions.

A glance at Figures 2.3 and 2.5 proves that the Kyoto protocol was ineffective. The total global CO_2 emissions not only did not decline to the goal of 21,600 million tons, as agreed in the treaty, but almost doubled to 39,100 million tons in 2025. The main reason for this failure is the stipulation that the highly populous developing countries are not obligated to participate in the emissions reductions. The result of the Kyoto agreement is that the CO_2 emissions of several industrialized countries slightly decreased (especially in the EU, and Japan) during the period 2005–2015, but not as much as intended by the protocol. However, the growth of energy consumption and fossil fuel combustion in most developing and populous countries by far offset this reduction. An undesirable effect for the signatories of the Kyoto protocol is that, while some industries invested in energy-saving technologies and reduced their emissions, several energy-intensive industries (e.g., cement, aluminum, steel, fertilizers) found the high energy prices anti-competitive, had to close their European plants, and relocate to countries with cheaper energy, a phenomenon referred to as *carbon leakage* [34].

2.3.4 The 2015 Paris Agreement

After realizing that the Kyoto protocol was a failure, the UNFCCC substituted the protocol with the Paris agreement, which attempts to include all nations – developed and developing – in the reduction of the GHG emissions. In December 2015 the Conference of Parties convened in Paris, France, where the parties reached another agreement, the main elements of which are [35]:

1. Reaffirm the goal of limiting the global temperature increase below 2°C and urge efforts to limit the increase to 1.5°C.
2. Ask for binding commitments by all parties to make *nationally determined contributions* (NDCs) for CO_2 emissions reductions.
3. Commit all signatories to report regularly on their emissions, the progress made in implementing and achieving their NDCs, and to undergo international review.
4. Commit all countries to submit new NDCs every five years, with the clear expectation that they will represent progress relative to previous NDCs.
5. Reaffirm the binding obligations of developed countries to support the efforts of developing countries.
6. Encourage voluntary actions by developing countries.
7. Extend the goal of mobilizing $100 billion annually in support of the agreement by 2020–through 2025, with a new, higher goal to be set for the period after 2025.
8. Extend a mechanism to address "loss and damage" resulting from climate change.
9. Eliminate double counting in reporting international emissions trading.

Of the 197 nations represented in the UN, only Yemen and Libya did not ratify this agreement. The USA was an original signatory, withdrew from the agreement in 2017, rejoined in 2021 and withdrew again, in 2025, following the political party in the Presidency.

A close look at the Paris agreement proves that its mechanism is entirely voluntary since the UN and the associated COPs do not have enforcing power. It stops short of imposing limits on CO_2 emissions and does not include sanctions for countries that violate their promised "contributions." The commitments by the signatory countries are "to report," "to monitor," "to submit new NDCs," "to encourage," "to ask for commitments," etc. In the entire text of the agreement, there is not a single nation or region obligated to actually curb their CO_2 emissions or to do anything that specifically and meaningfully addresses their emissions other than offering new NDCs, which may or may not be significant.

When it comes to actually reducing the national and global emissions, the Paris agreement makes use of the word "should," which implies advisement and consideration, instead of the word "shall," which denotes a legal obligation. It is characteristic of the lack of firm commitment and concrete action by all the signatories of the Paris agreement that, ten years after its signing essentially no progress has been made in actually curbing the global CO_2 emissions.

Table 2.3 shows the growth of the anthropogenic CO_2 emissions as well as several other key variables associated with energy use and emissions in the ten years following the signing of the Paris agreement. One may easily see in this Table that not only have CO_2 emissions significantly increased since the adoption of the Paris agreement

TABLE 2.3
Key Variables Associated with CO_2 Emissions in the Years 2015 and 2025

	2015	2025	% Increase
Global CO_2 emissions, million tons	35,209	39,100	11.1
Population, millions	7,380	8,230	11.5
Total Primary Energy, EJ	556	625	12.4
Total Electricity, TWh	20,253	32,400	60.0
Total Coal, million tons	7,860	8,900	13.2

but that other associated variables do not show any signs of reduction. Even the use of coal – a fuel that produces a great deal more CO_2 per unit heat generated and is of first priority to be eliminated – has increased by more than 13%. Indeed, recent investments in coal mining have been very profitable for investors in the developing world [36]. As a result of the higher coal burning and increased anthropogenic emissions, the World Meteorological Office (a UN agency) reported in October 2025 that, instead of slowing, the annual growth rates of CO_2 atmospheric accumulation have tripled since the 1960s, accelerating from 0.8 ppm per year to 2.4 ppm; and that the 2023–2024 global growth of the CO_2 concentration was 3.5 ppm, the highest increase since measurements started in 1957 [37]. It appears that, ten years after the Paris Agreement, there are no tangible results to support its underlying promise, to actually reduce the atmospheric CO_2 concentration.

In summary, the Paris agreement "has no teeth" to force national governments to reduce their CO_2 emissions. A few nations – chief among them the EU, Japan, New Zealand, Australia, and, partly, USA and Korea – have done so but their effort has been thwarted by others that continue their old, convenient, and less costly practices. The pledges for "national contributions" are entirely voluntary with no consequences for those that do not meet their "contributions." The displeasure for lack of concrete action by many national governments is alluded in several UN documents, as for example the IEA [38] *roadmap for Net-Zero emissions by 2050*, which complains that: "However, most pledges are not yet underpinned by near-term policies and measures …." When it comes to other more urgent problems, and especially national economic problems, governments do not have the political will to take immediate climate action and "kick the can down the road" for their successors to pick.

2.4 ENVIRONMENTAL REMEDIATION SUCCESSES OF THE RECENT PAST

Global warming is not the first environmental problem the world experienced. Acid rain, lead in the atmosphere, and the ozone hole in the Antarctic were global environmental problems that troubled the human society in the late 1900s but were successfully solved and have almost been forgotten. The solutions were achieved after concerted efforts by scientists, engineers, and governments. This section examines these environmental successes and, especially, how these environmental problems were solved.

2.4.1 Acid Rain

When fossil fuels are extracted from the earth's interior, they also contain small quantities of sulfur compounds. When the fuels are burned to form CO_2, the sulfur produces sulfur oxide, SO_2, which is released in the atmosphere with the other exhaust gases. In the atmosphere, SO_2 combines with moisture to form hypo-sulfuric acid (H_2SO_3), a relatively mild acid. If combined with ozone, SO_2 forms the much stronger sulfuric acid (H_2SO_4). In a similar way, nitrogen oxides (also in the combustion products) produce the weaker nitrous acid (HNO_2) and the much stronger nitric acid (HNO_3). All these acids are formed within the atmosphere inside tiny droplets or on the side of very fine particles, which are called aerosols. The sizes of aerosols are in the submicron range – one millionth of a meter – and this means that aerosols remain airborne for weeks or months and follow the predominant air currents that transport them over distances of hundreds of kilometers. However, during rain and snow, the aerosols combine with the larger rain drops and snowflakes and precipitate on the ground as *acid rain* or *acid precipitation*. The rain runoff, which contains higher concentration of the acids, eventually feeds the water of rivers and lakes and makes them acidic.

Scientists measure how strong acids and bases are, by a property called *potentiel Hydrogen*, or *pH* for short. The pH of pure water is 7. Acids have lower than 7 pH and bases have pH higher than 7. The pH of vinegar is between 2 and 3, and the pH of the much stronger sulfuric acid is less than 1. The normal pH level of natural freshwater is in the range 6.8–7.6 and the pH of oceans is close to 8. Most fish and several aquatic organisms do not thrive outside this natural range. The acidic rain runoff lowers significantly the pH in rivers and lakes, with the result that several aquatic species languish or disappear.

I personally experienced the effects of acid rain during my first year at Oxford University. While jogging under a light rain (a very common weather occurrence in England) in the park, I noticed that one of my eyes was irritated after a droplet got into it. I attributed the irritation to a dust particle and turned my face up, with both eyes open, so that the rain would cleanse the foreign object and soothe the irritation. To my astonishment, as the rain drops entered my eyes, the irritation became a strong burning sensation that only subsided after a stream of tears and washing with clean tap water. That was the end of my jogging-under-the-rain days … We learned from the newspapers on the following day that the pH of that rain was 3.4 and this was not as bad as in other places: a storm in Scotland in 1974 rained with pH 2.4. The pH of a rainfall in Kane, Pennsylvania, on September 19, 1978, was 2.32. During the entire 1970s, the pH of 80% of drizzles in Holland was less than 3.5, and oftentimes as low as 2.5. In short, the sky was discharging an acid, sometimes stronger than vinegar [39].

As the runoff acidic water mixed with the natural water systems the pH of rivers and lakes dropped significantly in the 1970s, well below the safe levels for aquatic species. Several species disappeared from the lakes. The low pH and lack of nutrients had catastrophic consequences on the aquatic populations in the majority of rivers and lakes in northern Europe and North America. In several Norwegian rivers salmon did not reproduce for years and became almost extinct. The fish disappeared entirely from 190 lakes in the Adirondacks region of the New York State and Canada, and from more than 2,000 lakes in southern Norway. Acid rain destroyed several aquatic ecosystems and became the preeminent global environmental threat of the 1960s and 1970s [40, 41].

What accentuated the environmental problems of acid deposition is that often the production of SO_2 and other oxides occurred in other states and not in the places where the acid rain fell. The acid-laden aerosols were carried by the predominant air currents over state and national boundaries and the acid rain affected neighboring states and nations. SO_2 aerosols produced in Pennsylvania, Ohio, and Michigan in the USA were carried north and affected the lakes in the northern New York State, Vermont, and the Canadian provinces of Ontario and Quebec. Similarly, acid oxides produced in the industrial Ruhr region of Germany affected the aquatic environments of Holland and Scandinavia. Environmental pollution does not respect national boundaries and becomes an international issue.

The engineering community developed an effective solution to the acid rain problem, the *in situ* (in place) removal of SO_2 using the *flue gas desulphurization* method. The method removes the SO_2 from the combustion gases by *scrubbers* before this gas is discharged to the atmosphere. Effectively, the gaseous combustion products are "showered" with a limestone-water mixture that contains calcium hydroxide, $Ca(OH)_2$. The SO_2 in the flue gases reacts with this basic solution and forms calcium sulfate, $CaSO_3$, a solid substance that precipitates in the bottom of the scrubber and is removed as sludge. $CaSO_3$ is chemically inert and not a pollutant. It is typically disposed of at the site of its production. Most of the nitrogen-oxide emissions are also captured by this chemical method.

When the governments decided in the early 1980s to react to the acid rain problem, this effective and relatively inexpensive engineering solution was available. The Environmental Protection Agency of USA introduced a *cap and trade* program that set upper limits (caps) for the emissions of SO_2 at electric power plants and refineries [41]. Canada adopted a range of complementary policies to gradually limit its own emissions [42]. The EU countries also took measures for the reduction of the SO_2 and NOX emissions: the Cooperative Program for Monitoring and Evaluation of the Long-range Transmission of Air Pollutants in Europe, which started in 1980, achieved a 75% decrease of the SO_2 emissions between 1980 and 2005 [43]. Japan and other countries in Asia, followed suit [44].

With the concerted national efforts, the acid precipitation reduction programs have been immensely successful. Figure 2.7 shows the dramatic drop of the SO_2 emissions in the USA and Europe as a result of the implementation of acid precipitation reduction measures. It is noted that, by 2025 the SO_2 emissions in the USA and Europe were under 1,700 tons, a 95% reduction from the 1970s levels. As a result, most of the environment and the affected aquatic ecosystems have returned to their previous, almost pristine, conditions. Acid deposition is not considered an environmental problem in the 21st century.

In addition to scrubbers the engineering community devised other pathways for the reduction of SO_2 and NOX emissions, such as using Fluidized Bed Reactors,

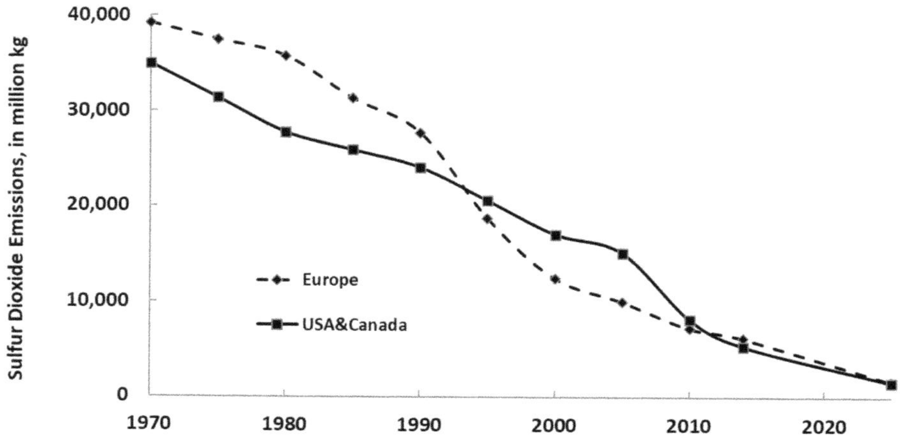

FIGURE 2.7 Reduction of SO_2 emissions in USA and Europe between the years 1970 and 2025.

blending high-sulfur coal with low-sulfur coal, substituting coal as fuel with natural gas, and retiring old and inefficient electricity generation units.

It is significant that not only the long-term goals of the acid reduction programs were achieved, but also that the costs of the abatement implementations to the businesses and the consumers were lower than the original predictions. It is estimated that the total cost of the SO_2 emissions reduction implementation strategies in the USA was in the range 1-2 billion US$, approximately one fourth of the original estimates by the coal industry and the electricity generation corporations. A large fraction of the sulfur abatement costs was recovered by the increased efficiency of the new equipment in the power plants. It is also important that the sulfur abatement did not cause any disruptions of the electric power generation and no inconvenience to electricity consumers. All in all, with the concerted and coordinated efforts of governments and the international scientific community, the harmful effects of acid precipitation have been eliminated at a very small cost to the electricity generation industry and the consumers.

2.4.2 Lead Abatement

Gasoline and diesel fuel are mixtures of liquid hydrocarbons derived from crude oil after refinement. Both fuels are mixed with air and burn in the cylinders of internal combustion engines to produce motive power. The gasoline-air mixture in the cylinders is designed to ignite by a spark. However, a few of the hydrocarbons in the gasoline mixture reach their ignition point, auto-ignite, and release heat prematurely, before the spark is triggered. This causes the "knocking" problem in gasoline engines, where premature ignition introduces vibrations, lower power, and engine damage.

Auto-ignition in gasoline engines may be prevented by chemical additives, the most common of which in the past was tetra-ethyl lead, $Pb(C_2H_5)_4$. When added to the gasoline, tetra-ethyl lead prevents premature combustion, knocking, and engine damage; and enables the engine to run smoothly. The use of this chemical was widely adopted by the refining and automobile industries as an "anti-knock" additive to the gasoline in the early part of the 20th century. However, the $Pb(C_2H_5)_4$ burns together with the fuel. Its combustion releases lead oxides to the atmosphere, primarily PbO and Pb_2O, as well as atomic Pb. Humans and animals breathe or ingest these chemical compounds, with the result that lead enters their bodies. A second pathway for the human absorption of lead is via the lead-based pigments and the sweetener lead acetate, which is also known as "sugar of lead." Pigments with lead base have been used for centuries to produce bright colors of yellow ($PbCrO_4$), red (Pb_3O_4), and white ($PbCO_3$). Lead-based pigments were used, worldwide, for the manufacturing of paints for houses and commercial buildings as late as the 1980s.

A number of toxicological studies in the 1960s have proven that chemicals containing lead are harmful to the health of the humans, because lead compounds affect the synapses of brain cells, especially those of children. Prolonged exposure to lead has been proven to cause mental retardation and brain disorders to humans, including dementia, while the ingestion of large quantities of lead will cause death (lead poisoning).

Because of its severe health effects, national regulations were enacted in the 1970s to phase out the use of the $Pb(C_2H_5)_4$ from gasoline additives. Lead-based chemicals were also phased out from paints and other common materials as well as from all the equipment and additives used in the food industry. In the USA and the countries of the EU the sale of leaded gasoline for automobiles has been completely banned since the early 1990s (limited amounts of leaded gasoline are still allowed for marine engines, racing cars, and a few types of agricultural equipment). By 2020, all other countries have adopted similar regulations and tetra-ethyl lead has been replaced by other chemical additives. The use of leaded gasoline for road vehicles has been globally banned, and the United Nations Environment Programme declared the official end of leaded gasoline use for cars on August 30, 2021.

The restrictions and eventual ban on leaded gasoline, as well as lead-based paint caused the average concentration of lead in the atmospheric air to drop precipitously. Figure 2.8 depicts the dramatic decrease of the average lead concentration in the air in the USA [45]. It is observed that, between 1980 and 2013, the ambient lead concentration decreased by a factor higher than 12. Blood tests show that the lower ambient lead resulted in the lowering of lead in the human bloodstream, especially in children, and this implies lesser neurological disorders and improvement of the quality of life. As with the acid rain, the elimination of large quantities of lead from the environment was achieved with prudent regulations, technology that became available by the scientists and engineers, and with minimal cost and inconvenience for the consumers and industry.

2.4.3 Ozone Hole and the Montreal Protocol

Ozone gas, O_3, is composed of three oxygen atoms and is very reactive. Ozone gas in the lower atmosphere is a pollutant and toxic to humans but in the upper atmosphere it is vital for the preservation of life on earth. Because of its very high reactivity,

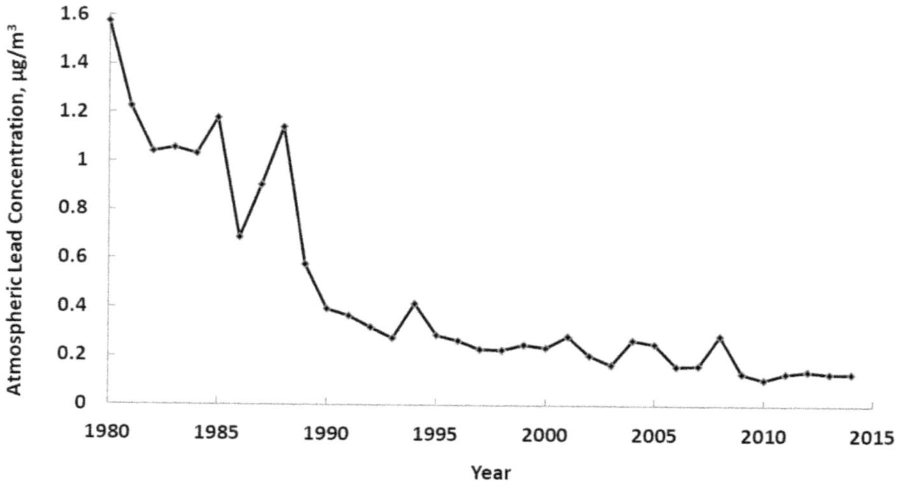

FIGURE 2.8 The average lead concentration of air in the USA diminished in the period 1980–2013.

high ozone concentration in the lower atmosphere (the biosphere) causes throat and lung irritation, asthma, and emphysema to humans. However, ozone in the stratosphere – the part of the atmosphere between 10 and 50 km – is necessary for the absorption of the high-energy ultraviolet (UV) sunlight that can destroy life forms. Without ozone in the stratosphere, the sun's UV radiation would effectively sterilize the earth's surface and would kill living organisms. Effectively, the ozone gas in the stratosphere provides a "sunscreen" that eliminates all of the most energetic UV-c radiation, about 9% of the UV-b radiation and 5% of the less harmful UV-a radiation. Excessive amounts of these types of radiation cause severe sunburns that may lead to skin cancers and eye damage. In short, ozone in the lower atmosphere is detrimental to human health, but its presence in adequate concentrations in the stratosphere is necessary for the preservation of life on earth. Figure 2.9 shows the concentration profile of ozone in the atmosphere as a function of the height. It is observed that only traces of the gas are present near the surface of the earth, where the gas is harmful to humans. The concentration of ozone increases significantly in the stratosphere, where it absorbs the damaging ultra-violet solar radiation, and reaches a maximum close to 8 ppm.

Atmospheric monitoring studies in 1985 showed that the stratospheric ozone concentration above Antarctica had fallen significantly [46]. The measurements confirmed a previous analytical study (of 1974), which concluded that the continuously increasing atmospheric concentration of chloro-fluoro-carbon (CFC) refrigerants – relatively inert chemicals that may linger in the atmosphere for 40–150 years – generates the chemically active chlorine atoms that combine with the atmospheric ozone in the troposphere and cause its depletion [47]. The partial depletion of ozone in the arctic regions was dubbed "ozone hole." Actually, it is not a hole at all but a very large area above the Antarctica, where the concentration of ozone is

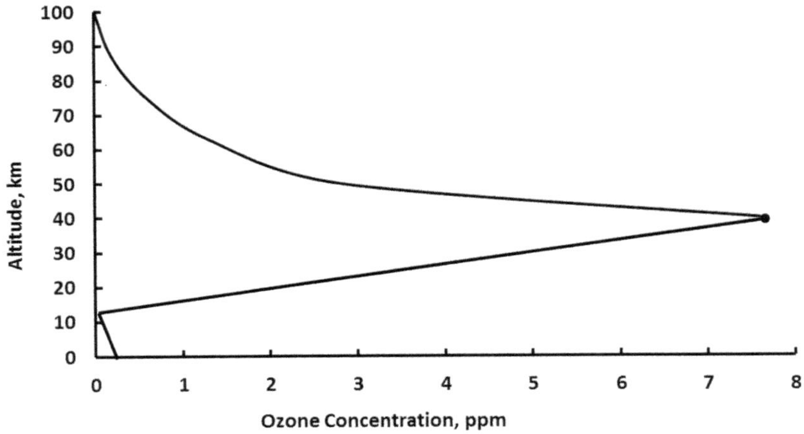

FIGURE 2.9 Ozone concentration in the atmosphere.

significantly less than the global average concentration. As a result of the ozone hole, higher fractions of harmful UV radiation penetrate the upper atmosphere and reach the surface of Antarctica.

The appearance and growth of the "ozone hole" in the Antarctica is alarming, because its extension to other, more densely populated areas of the planet has the potential to cause disastrous health effects to the human population. An initial international conference was held in Vienna (1985), where it was decided that refrigerants, which contribute to the ozone destruction, should be gradually banned. The *Montreal Protocol* (1987) ratified this decision and set the goal that by 1996 there would be zero production of ozone-depleting CFCs.

This resolute international pact for the elimination of the harmful CFC's was successful: new refrigerants, the hydro-fluoro-carbons (HFCs, such as R-134a, R-510, etc.) were developed between 1987 and 1996. They substituted the CFCs in refrigerators and air-conditioners. The atmospheric concentration of the CFCs gradually diminished, the ozone concentration stabilized and started to increase after 2001. Climatological projections show that the ozone layer in Antarctica will return to the before 1980 levels sometime between 2050 and 2070 [48].

2.4.4 Hydrofluorocarbons and the Kigali Agreement

A short time after the adoption of HFCs as new refrigerants, scientific studies showed that these chemicals are very potent GHGs, with some of them (e.g., R-142b and R-134a) being close to five thousand times more potent than CO_2. This led to the signing of the *Kigali agreement* [49] for the elimination of the HFC refrigerants in October 2016. The agreement calls for the total substitution of the highly potent HFCs with hydro-fluoro-olefins (HFOs) which are benign to the ozone layer and are not potent GHGs by 2050 [50, 51]. The substitution of the HFCs started in the developed countries in 2019. The developing countries have until 2028 to acquire the technology and implement this substitution. At the time of its adoption, it was estimated that the effect of the elimination of the HFCs by 2050 will reduce the potential global warming at the end of the 21st century by 0.5°C [52].

Even though the Kigali agreement has not been in full force and its global effects have not been assessed, I consider that it will be fully successful, and I do not hesitate to include it in the list of "environmental successes" for the following reasons:

1. The scientific community has available ready substitutes for the HFCs, the HFOs. No new technology is required for the adoption of the Kigali agreement.
2. The manufacturing of HFOs is inexpensive and the engineering modifications to the refrigeration systems are modest and may be accomplished at relatively low cost.
3. Because the adoption of the HFO refrigerants will not have a disruptive effect on the economies of the countries that have implemented it, the political community has accepted this viable solution and proceeded with national mandates.

2.4.5 Characteristics of the Four Environmental Successes

The abatement of lead, the elimination of acid rain, and the substantial recovery of the Antarctic ozone layer, are global environmental success stories. The expected elimination of the HFCs will soon become another success. In all four environmental success stories the threats to the ecosystems were identified by scientists using accurate scientific data. The international scientific community analyzed the data, drew its conclusions about the severity of the threats, and proposed sound remediation solutions to the political communities and the regulatory branches of governments. The causes of the environmental threats were eliminated by concerted international effort; the global industries were given alternative solutions and a firm timeframe to significantly reduce or eliminate the offending materials; the healing processes for the environment started promptly; and with continuous monitoring by the regulatory agencies the results of these international efforts have proven to be successful.

The scientific and technical communities played a significant role in all the environmental mitigation success stories. Sulfur scrubbers were developed for coal power plants and refineries to eliminate the SO_2 emissions and acid rain. Lead-free anti-knock chemicals were developed and massively produced to eliminate the lead oxide emissions. Refrigerants that do not attack the atmospheric ozone were developed and substituted the ozone-destroying CFCs. In all these success cases, the technical community proposed scientifically sound and affordable solutions to alleviate the environmental threats. Despite an initial resistance from the industrial stakeholders (coal power plant operators, gasoline producers, refrigerator manufacturers, etc.) the regulators relied on the scientific data and the proposed solutions to require that the relevant industries follow the technical solutions. The result in all cases is that scientists, technologists, politicians, and regulators found common ground with industries to adopt sound scientific and engineering measures that alleviated the three environmental threats in a relatively short time and rather inexpensively.

The elimination of the HFOs, following the Kigali agreement, adopted a similar strategy and is expected to be successful.

Another characteristic of these global environmental triumphs is that the general public understood the risks. The international press followed the scientific community and presented the problems as indisputable conclusions based on scientific data. No one muddled the waters by asking questions such as: "do you believe in acid rain?" or "do you believe that lead is harmful to your children?" The general public was presented with the scientific findings to understand the causes, risks, and ecological dangers of the environmental problems. Following the public sentiment, political parties and governments had no other option but to adopt the solutions offered by the scientific community, and to enforce regulations directing the industries to stop emitting the offending pollutants.

2.5 UNIQUENESS OF THE GLOBAL CLIMATE CHANGE PROBLEM

Four factors substantially contributed to the successful solutions of the past environmental problems:

1. The technology for all solutions was more or less known and easy to develop. In all cases the culprits were chemicals that were substituted with different, non-offending materials.
2. The proposed solutions were confined to a particular industry and were not expected to have any detrimental effects on the entire global economy.
3. The costs of the solutions to the affected industries were relatively low.
4. There were no public queries or requests for monetary "penalties," "reparations," "victim compensations" or "loss and damage compensation." The affected industries proceeded to make changes without any legal obligations or legal threats.

The obvious solution to the GCC problem is to substantially reduce the GHGs, most importantly to reduce the global CO_2 emissions. However, a simple and relatively inexpensive solution to this problem does not exist, as it will become apparent in Chapter 3. Drastic CO_2 emissions reduction requires drastic fossil fuel reduction that (at the present state of technology) will cause substantially higher energy costs, and very likely energy shortages. This will have a dramatically adverse effect on national economies that depend on inexpensive and reliable energy. The gradual transition to renewable energy sources is possible but requires a great deal of capital for infrastructure and global cooperation.

A major disadvantage of the Paris agreement to solve the GCC problem is that it is not binding for nations. The UN, which oversees the effects of the agreement, does not have enforcement power. The nationally determined contributions are just that: contributions. If member nations do not meet them there is only shame on them and no sanctions. Moreover, the contributions of most developing nations are "subject to the availability of capital." Unless substantial loans (or preferably grants) are received from the World Bank or other countries, the "contributions" of many developing nations will not materialize. Our experience with the climate fund since 2010 is that there are few grants, and that substantial loans are difficult to come. On the other side, the governments of the developed nations are reluctant to underwrite the energy transition, not only in the developing world but also in their own countries.

A factor that delayed the realization of the enormity of the GCC problem and still delays the urgency to act is the framing of the problem not in scientific terms but as one of each individual's beliefs. The general public is frequently bombarded with "news" that include opinions of individuals who may be unfamiliar with basic sciences, climate, and statistics, but freely pontificate about their "beliefs" on global warming and GCC. The tendency of the information media to create a "balance" in every story complicates matters by deliberately bombarding the public with the opinions of unqualified individuals, both climate activists and deniers.

We all have to accept that the GCC is a scientific question to be proven or disproven by reliable data and expert scientists. The beliefs of poorly educated individuals, activists, politicians, and "celebrities" do not carry any weight in such matters. Let us not forget that, only a few centuries ago, most individuals believed that the earth is flat (and supported by four pillars!) and that the sun rotates around the earth. Science has proven these beliefs wrong, and we are now able to circumnavigate the globe. Scientists resolve scientific questions, such as the GCC, by obtaining accurate data and drawing logical conclusions based on the data. The "beliefs" of individuals are entirely irrelevant in matters of science. It is rather unfortunate that, in the case of GCC, the deniers have muddled the waters by bringing the "beliefs" of under-educated laymen in the debate, while the avowers introduced a highly exaggerated sense of urgency, sometimes using yet-unqualified teenage girls (Greta, Adegbile, and others) as their spokespeople in this highly complex scientific matter, heavily laden with uncertainty.

By its global nature the CO_2 emissions problem requires the cooperation and adherence to the guidelines and regulations by all the nations on the planet. More than ten years after the introduction of the Paris agreement, the progress to achieve the NDCs has proven to be inadequate, and it is very likely that the agreement will be swapped with another in the near future that will be more binding. Before this happens, the international community will have to agree on what is the timescale and what is a "fair" manner to allocate national CO_2 emissions reductions by answering questions such as:

1. What should the global limit of total CO_2 emissions be and when must it be achieved?
2. Should the OECD countries, which typically produce the highest CO_2 emissions *per capita* but have almost zero population growth, be allocated the highest *per capita* CO_2 emission reductions?
3. Is it fair that some nations have higher CO_2 emissions *per capita* than others?
4. Should countries with historically fast-growing populations be allocated the same *per capita* CO_2 emissions allowances as countries with stagnant populations?
5. How will the populations of countries like USA and Canada (and China now), which have exceptionally high emissions, adapt to a CO_2 reduction mandate?

6. Should landlocked countries in colder regions – Switzerland, Tajikistan, Mongolia, and similar countries, that are not threatened by GCC participate in a global attempt for CO_2 emission reductions?
7. Should the developed countries pay for "loss and damage?"
8. Should fossil fuel producing corporations/companies pay for "loss and damage?"

It is apparent from these questions that the GCC problem has legal, ethical, moral, and technological implications that make it unique in the history of our world.

2.6 MYTHS AND REALITY RELATED TO GLOBAL CLIMATE CHANGE

Because so many opinions and "beliefs" are expressed daily in information media, several myths have been promulgated about GCC, its effects, and its possible solution. A handful of such myths are exposed here, and a response is offered, the reality, which is based on science and the available scientific data. The quotes in italics in this section have been lifted from statements made over the years by journalists, activists, government officials, and other influencers of public opinion. The source of the quotes is not given to avoid embarrassment. However, we all remember hearing or reading similar statements that lack scientific rigor and were made to animate or appease the general public.

Myth # 1: There is no global warming; the global temperature variability is due to solar activity.
Reality: While solar activity affects the total radiation from the sun and the earth's temperature, solar activity is periodic with a timescale of 11 years. The data in Figure 2.4 (since 1910) span more than ten solar cycles and clearly show that the long-term average temperature of the biosphere continuously increases at timescales longer than 11 years. Since the increase continues after ten solar cycles, it is reasonable to conclude that something other than solar activity (the GHGs) is the cause of this long-term temperature increase.

Myth # 2: The most recent three-year temperature data show that there is no global warming.
Reality: While there have been short periods when the earth's average temperature did not increase (or even decreased!) these years were followed by longer periods when the temperature significantly increased. The long-term increasing trend was verified as shown in Figure 2.4. The global climate is a long-term characteristic and only long-term temperature data should be used to draw conclusions on GCC.

Myth # 3: How can we have global warming? Here is a snowball I picked up from the streets of Washington DC today!
Reality: Here is someone who confuses weather with climate. Snow in Washington DC during a winter day does not mean that the climate – the long-term-average weather – is cooling or that there is no increase of the

average yearly temperature. Even if the average annual temperature in Washington DC increased by 5 or 10°C, there would still be a few winter days with snowfall and, even, snow blizzards.

Myth # 4: The average global temperature data is inaccurate. We need several more decades of data to draw accurate conclusions.

While each of the yearly average global temperature data is laden with uncertainty, statistical analyses of the entire set of data since 1900 have shown that the long-term, increasing trend is correct and significant. There is more than 99% statistical confidence that the atmospheric temperature increased by 0.55 to 0.67°C during the 20th century [53] and that this trend continues in the 21st century with at least another 0.5 °C temperature rise. In addition, there is the cause-and-effect relationship between GHGs and earth's temperature. The cause (GHGs cause warming) is supported, actually verified, by the observed effect (higher temperatures since 1900) with the conclusion that the continuously increasing GHG emissions will result in even higher temperatures in the future.

Myth # 5: Hurricane Sandy that flooded New York is due to global warming.

Reality: The implication of this statement is that if we reduce the CO_2 build-up in the atmosphere, hurricanes, monsoons, and similar disasters will stop. While some computer models predict the intensification of tropical storms with increasing CO_2 – including cyclones and hurricanes – the storms are weather phenomena, and, by themselves, do not signify climate change. Hurricanes and tropical storms have been recorded for millennia, before any evidence of global warming, and will continue to occur even if we reversed the warming trend.

Myth # 6: All we need to do in the USA to remedy GCC is to levy a small tax on the coal industry.

Reality: Given the current global structure of the electricity production industry, a small tax on the USA coal will not significantly reduce the global CO_2 emissions. A significantly high tax on coal in the USA alone will make wind and solar energy more competitive but will also increase the electricity prices and reduce the electricity generation reliability. A coal tax (or carbon tax) in a single country will hardly make a dent on the global CO_2 emissions if other countries do not take similar measures to reduce their own CO_2 emissions. The GCC and CO_2 emissions are global problems, and their solution must involve all the countries on the planet. The failure of the Kyoto protocol proved that curtailing the CO_2 emissions in a single country or a small group of countries will not mitigate the GCC problem.

Myth # 7: We should not rush to take any measures about CO_2 emissions because scientists still disagree on global warming.

Reality: The overwhelming majority of recognized experts on climate agree that the earth's temperature is increasing and have urged the political establishment to act. The few "scientists" that have different opinions, either are unfamiliar with climate models or are very closely connected to the CO_2 emissions industry to have a conflict of interest, and their opinions are strongly biased.

Myth #8: Stabilizing the world's climate will require high-income countries to reduce their emissions by 60%–90% from the 2006 levels by 2030.
Reality: In the absence of technological breakthroughs with renewable energy sources, such drastic reductions in the near future will ruin the economies of the "high-income countries," and their citizens will be unable to maintain their "high-income" status. Citizens of the developed nations will never condone unilateral CO_2 emissions reductions of this magnitude without commensurate concessions and reductions from developing countries. GCC is a global threat that must be wrestled by all countries, wealthy and poor.

Myth #9: The chance that there will be any permanent ice left in the Arctic after 2022 is essentially zero.
This is a statement that originated from a lecture by a climate scientist, was misquoted in a journal, and then made the rounds in the information media. The year 2022 came and went and the Arctic ice is still there. The original statement by the scientist referred to floating ice, but in 2025 there is a great deal of floating ice in the Arctic Ocean [54]. The failure of such "predictions" only emphasizes the high uncertainty associated with all climate models and that scientists must be very careful with any definitive "conclusions."

Myth #10: Climate change will wipe out all humanity unless fossil fuel use is ended over the next five years (by 2023).
This quotation by a climate activist originated from the statement of myth #9. It is intended to alarm the general population but is not based on any scientific data or other evidence. Headlines like this attract attention, frighten the general public, and add to the urgency of reducing the combustion of fossil fuels and the CO_2 emissions. However, most of them are either misguided opinions of activists (and journalists who do not dig deeper to find the truth) or snippets lifted from scientific articles (without any context) that do not reflect the more intricate conclusions of the studies. The worse consequence of such doomsday predictions is that, when they prove to be inaccurate, they also fuel doubts about the validity of other *bona fide* climate data and studies.

NOTES

1 Fourier, J.B.J., 1824, Remarques générales sur les températures du globe terrestre et des espaces planétaires, *Ann. Chim. Phys.* (Paris) 2nd ser., 27, 136–167.
2 Tyndall, J., 1861, On the Absorption and Radiation of Heat by Gases and Vapours, and on the Physical Connection of Radiation, Absorption and Conduction, *Phil. Mag. ser. 4*, 22. 169–194.
3 Arrhenius S., 1896, On the Influence of Carbonic Acid in the Air upon the Temperature of the Ground, *Philos. Magazine and J. of Science*, 5th series, 41, 237–276.
4 Arrhenius S., 1908, *Worlds in the Making – The Evolution of the Universe,* (transl. H. Borns) Harper & Brothers, New York.
5 Sherwood, S.C. *et al.* (25 authors), 2020, An Assessment of Earth's Climate Sensitivity Using Multiple Lines of Evidence, *Reviews of Geophysics*, 58, e2019RG000678. https://doi.org/10.1029/2019RG000678.

6 Data from https://github.com/owid/co2-data (accessed May 2026).
7 Michaelides, E.E., 2017, *Energy, the Environment, and Sustainability,* CRC Press, Boca Raton FL.
8 IPCC, 2007, Climate Change 2007: Synthesis Report. Contribution of Working Groups I, II and III to the Fourth Assessment Report of the Intergovernmental Panel on Climate Change, IPCC, Geneva, Switzerland.
9 International Energy Agency, 2019, CO2 Emissions from Fuel Combustion – Overview, IEA Publications, Chirat, Paris.
10 National Oceanic and Atmospheric Administration data from https://gml.noaa.gov/ccgg/trends/.
11 The uneven thermal interactions of the earth's surface also cause the winds, both local and planetary.
12 Data from https://www.ncei.noaa.gov/access/monitoring/climate-at-a-glance
13 Mann M., 2014, False Hope – Earth Will Cross the Climate Danger Threshold by 2036, *Scientific American*, 310 (4), April.
14 Kotz M., Levermann A., Wenz, L., 2024, The economic commitment of climate change, *Nature*, 628, 551–557.
15 IPCC, 2022, *Climate Change 2022: Mitigation of Climate Change* (report available at the IPCC website: www.ipcc.ch).
16 Interactive map: How has local sea level in the United States changed over time? https://www.climate.gov/news-features.
17 Sweet, W. V., et al., 2022, *Global and regional sea level rise scenarios for the United States,* USGS Report 70229139, https://www.usgs.gov/publications/global-and-regional-sea-level-rise-scenarios-united-states
18 Christy J., *et al.,* 2025, *A Critical Review of Impacts of Greenhouse Gas Emissions on the U.S. Climate,* US-DOE, Washington DC.
19 *Global and European Sea Level Rise,* 2025, European Environmental Agency Report. Available at https://www.eea.europa.eu/en/analysis/indicators/.
20 Rahmstorf S., 2007, A Semi-Empirical Approach to Projecting Future Sea-Level Rise," *Science* 315, 368–370.
21 Watts R.G., Morantine M., 1990, Rapid climatic change and the deep ocean, *Climatic Change*, 16 (1), 83–97.
22 Sarmiento S.L., Wofsy S.C., 1999, *A US Carbon Cycle Plan,* US Global Change Research Program, Washington DC.
23 Gattuso, J. P., and Hansson, L., eds, 2011, *Ocean Acidification: Background and History,* Oxford University Press.
24 Ballantyne A.P., Alden C.B., Miller J.B., Tans P.P., White J.W.C., 2012, Increase in observed net carbon dioxide uptake by land and oceans during the last 50 years, *Nature* 488, 70–72.
25 Clements, J.C., Sundin, J., Clark, T.D., Jutfelt, F., 2022, Meta-analysis reveals an extreme "decline effect" in the impacts of ocean acidification on fish behavior, *PLOS Biology*, 20(2), e3001511, https://doi.org/10.1371/journal.pbio.3001511
26 Zeng, Z., *et al.*, 2017, Climate mitigation from vegetation biophysical feedbacks during the past three decades, *Nature Climate Change*, 2399, https://doi.org/10.1038/nclimate3299
27 Cramer, W., et al., 2014, *Detection and Attribution of Observed Impacts,* in *Climate Change 2014: Impacts, Adaptation, and Vulnerability. Part A: Global and Sectoral Aspects* (C. B. Field *et al.* eds.), pp. 979–1037, Cambridge University Press.
28 Close to 60% of these contributions are in the form of loans rather than outright grants.
29 Crippa, M., *et al.*, 2024, GHG emissions of all world countries – 2024. Luxembourg: Publications Office of the European Union. doi:10.2760/4002897.
30 Burgueño Salas, E., 2025, Historic contributions to global warming worldwide 1851–2023, by country or region, *Statista*, August 7, 2025.
31 Data from https://data.worldbank.org/indicator/SP.POP.TOTL?locations=1W

32 It is characteristic of the rapidly increasing population of several developing countries that, when I worked out the same example in my book *Energy, the Environment and Sustainability,* with 2015 data, the CO_2 emissions increase was calculated as 1,694 million tons. By 2025, this number increased by 44%.
33 *Encyclopedia Britanica.*
34 Bollen J., Manders T., Tong P., 2003, The Kyoto Protocol and EU competitiveness, International Conference on Economic Modeling and Data Science (EcoMod2003), July, Istanbul, Turkey.
35 C2ES, 2015, *Outcomes of the U.N. climate change Conference in Paris, Center for Climate and Energy Solutions,* Arlington VA, December 2015.
36 Emont, J., 2025 The Billionaire Mining Magnate Who Bet Coal Had a Future –And Won Big, *The Wall Street Journal*, January 9.
37 WMO Greenhouse Gas Bulletin - No. 21, October 16, 2025, WMO, Geneva.
38 IEA, 2023, Net Zero Roadmap: A Global Pathway to Keep the 1.5°C Goal in Reach – 2023 Update, IEA publications, France.
39 Gates, D. M. *Energy and Ecology,* Sinauer, Sunderland Massachusetts, 1985.
40 Michaelides, E.E., 2012, *Alternative Energy Sources,* Springer, Heidelberg.
41 EPA, 2012, *SO_2 and NOx Emissions, Compliance, and Market Analyses.* Progress report-2012.
42 Environment Canada, 2012, *Canada-United States Air Quality Agreement.* Progress Report-2012.
43 Vestreng, V., Myhre, G., Fagerli, H., Reis, S., Tarrason, H., 2007, Twenty-five years of continuous sulphur dioxide emission reduction in Europe, *Atmospheric Chemistry and Physics*, 7, 3663–3681.
44 The more recent *Gothenburg Protocol,* which is promulgated by the UN and is global, is based on this success and is even more ambitious by covering more types of pollutants.
45 EPA, 2014, *National Trends in Lead Levels,* https://19january2021snapshot.epa.gov/air-trends/lead-trends_.html.
46 Farman J.C., Gardiner B.G., Shanklin J.D., 1985, Large losses of total ozone in Antarctica reveal seasonal ClOx/NOx interaction, *Nature*, 315, 207–210.
47 Molina, M.J., Rowland, F.S., 1974, Stratospheric sink for chlorofluoromethanes: chlorine atom-catalysed destruction of ozone, *Nature*, 249, 810–812.
48 Douglas, A.R., Newman, P.A., Solomon S., 2014, The Antarctic ozone hole: An update, *Physics Today*, 67, 42–48.
49 Kigali is the capital of Rwanda, a country in Central Africa.
50 Seidel, S., Ye, J., Andersen, S.O., Hillbrand, A., 2016, Not-in-kind alternatives to high Global warming HFC's, *Center for Climate and Energy Solutions,* Arlington VA, October 2016.
51 EIA, 2016, Kigali Amendment to the Montreal Protocol: a Crucial Step in the Fight Against Catastrophic Climate Change, Briefing to the 22nd Conference of the Parties (CoP22) to the United Nations Framework Convention on Climate Change (UNFCCC), November, Marrakech, Morocco.
52 U.S. EPA (U.S. Environmental Protection Agency). 2016. Inventory of U.S. greenhouse gas emissions and sinks: 1990–2014. EPA 430-R-16-002.
53 IPCC, 2014, Climate Change 2014: Impacts, Adaptation and Vulnerabilities, IPCC, Geneva, Switzerland.
54 https://nsidc.org/sea-ice-today

3 Climate Change Mitigation Approaches

With the continuously growing atmospheric CO_2 emissions that cause planetary warming; and threaten human communities and their economic activities, it is imperative that we study and implement some of the known and technologically feasible methods to reduce CO_2 emissions and mitigate the Global Climate Change (GCC). This chapter offers a succinct and critical survey on the key methods that have been proposed for the control and eventual reduction of the anthropogenic CO_2 emissions. Emphasis is given to energy efficiency, use of nuclear energy, and higher use of renewable energy – in particular wind and solar energy that are available everywhere on the planet – strategies that have been advocated in all the pathways and roadmaps to reduce CO_2 emissions. Detailed data on the conditions and requirements to decarbonize at three levels, a single home, a city, and an entire grid are also presented. The Chapter concludes with the effects electric vehicles may have on decarbonization efforts and the roadmaps presented by the agencies of the United Nations.

3.1 INCREASE PRICES TO REDUCE ENERGY CONSUMPTION (A CARBON TAX)

The law of demand and supply is a fundamental principle of economics. When the price of a good or a service increases, people find substitutes, and their demand falls. This law applies to fossil fuels and energy consumption, in general. If the price of fossil fuels were to significantly increase – e.g., by imposing a carbon tax – individuals will switch to other primary energy forms and develop conservation measures. They will use less of these fuels, and the final result will be less combustion products and less atmospheric CO_2 emissions.

The fossil fuel price increase will result in their substitution with less expensive energy forms (perhaps renewables) or with less energy consumption. While it is notionally possible to substitute fossil fuels with renewable energy forms, most practical energy systems are designed for a specific fuel, and substitution is either expensive, impossible, or takes a long time. My home heating system operates with natural gas. If the gas prices substantially increase and the increase appears to be permanent, I can substitute it with solar heating (or a heat pump that will consume electricity), but the new system requires significant investment expenses, and its installation will take a few months. For several industrial applications such a substitution is impossible. The cement manufacturing equipment operates at temperatures close to 1500°C (2700°F) and such high temperatures can only be achieved with fossil fuels or electricity, which is largely generated using fossil fuels. Substituting natural gas with solar heating for the production of cement is almost impossible, because the high temperatures cannot be attained and maintained with simple solar

 DOI: 10.1201/9781003674658-3

systems. The substitution of aviation fuels globally with renewable energy falls in the impossible category. Small quantities of aviation fuels may be substituted with biofuels, but this becomes an impossible proposition when applied globally, because not enough biofuels can be produced for the global aviation industry.

Taxing fossil fuels (carbon tax) makes energy more expensive and induces individuals and industries to consume less energy, two causes that have ripple effects in the economy. For individuals, spending more money to pay for necessary energy expenses (e.g., electricity for homes and gasoline for their cars to commute) implies that their disposable income is reduced and they have less money to buy other goods, leading to a contraction in the business activity. For industries, more expensive energy implies higher cost for their products. When the cost is passed on to the consumers at higher prices, the demand for the products will fall. This can be particularly distressing if the hike in the energy prices happens in one nation or a group of nations but is not global. The manufacturing products of the affected nations become more expensive, offering a competitive advantage to the industrial products of other nations that do not adopt carbon taxation. An example of this effect is the experience of the European Union (EU) industry after the Union adopted the Kyoto protocol (see Section 2.3.3). While the CO_2 emissions in the EU decreased (but not as much as originally intended) and some industries invested in energy-saving innovative technologies, the adoption of the Kyoto protocol incentivized energy-intensive industries (e.g., the cement and steel industries) to relocate to countries with cheaper energy, a phenomenon branded as *carbon leakage* [1]. In addition, the high cost of implementing the Kyoto protocol resulted in lower economic growth, estimated at 1 to 2% of the GDP in the early years of its adoption [2]. Almost thirty years later, the adoption of this protocol is principally blamed for the considerable drop of Europe's share of the global economic output – from 33% in 2005 to 23% in 2024 – while the USA and Japan essentially held on to their share of economic output [3]. Studies and policies that advocate a carbon tax justify it economically only if – apart from the direct energy costs for the manufacturing of goods and delivery of services – they take into account other intangible costs such as effects on human health, effects on the soil and agriculture, and environmental damage [4]. However, the latter costs are laden with very high uncertainty, and for the private and public corporations of most nations, they do not represent real costs that affect their bottom line.

A glance at the recent data proves that higher prices and lower energy consumption have always been associated with higher inflation and, often, with global economic slowdowns and contractions. Many of us (of a certain age) remember the quadrupling of the petroleum prices in the mid-1970s that was followed by the reduction of global energy consumption, and the reduction of global manufacturing activity. These effects ushered in the global recession and high inflation of the 1970s as well as the stagflation of the early 1980s, that adversely affected the global economy – and, particularly, the economies of the developing nations. Figure 3.1 demonstrates the high correlation between the annual growth of global GDP *per capita* (a measure of global economic progress) and the annual growth of energy consumption, both in percentages [5]. It is evident in Figure 3.1 that geopolitical events, which triggered global economic downturns, were also associated with lower energy consumption.

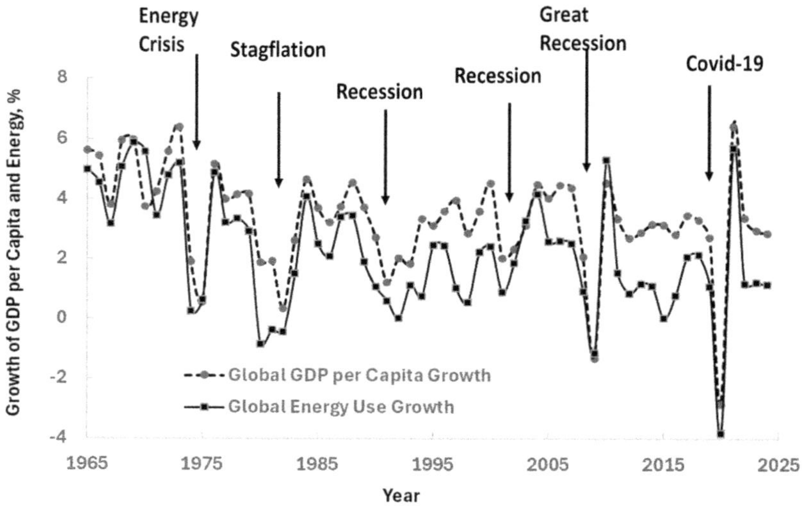

FIGURE 3.1 Annual growth rates of GDP *per capita* and global growth of energy consumption.

History has shown that negative GDP growth, which is a consequence of a global economic downturn, disproportionately affects the populations of developing nations. Poverty rates in the developing nations are higher, and when the nation's GDP falls, unemployment and poverty rates substantially rise, with the result that people cannot afford to buy necessities. The developing world needs more inexpensive energy to industrialize and catch up with the developed nations in the delivery of services, home construction, and the production of goods.

In summary, fossil fuel consumption and CO_2 emissions reduction will follow a carbon tax. However, the tax will have to be widely spread among all nations, so as to not put at a disadvantage the industries of the group of nations that adopt it. Also, the implementation of any carbon tax must be gradual and delicately controlled so as not to lead the global economy into an inflationary period or an economic downturn. When it comes to the developing world, a carbon tax is not a tool for prosperity. Developing nations need more affordable energy to combat income inequalities and poverty.

3.2 ENERGY EFFICIENCY AND CONSERVATION

The two terms are often used as synonyms to denote less energy use for a given action, service, or product manufacturing but in engineering practice they have slightly different connotations: higher efficiency is connected with equipment upgrade or replacement, which is accomplished with capital expenditure, while conservation is connected to actions of individuals and groups that save energy without any investment and may be coupled with a slight inconvenience. Replacing my old A/C unit with a new one of higher SEER [6] implies higher efficiency, while setting my thermostat to 18°C (65°F) and wearing an extra layer of clothing in the winter is an act of energy conservation [7].

Energy conservation is everybody's darling and has been touted by environmentalists since the 1960s as the best way to have a cleaner environment. We can decrease the amount of energy we consume and save money, by taking simple actions. Switching off lights when we leave a room, carpooling for the daily commute to the office with four neighbors, reducing the home temperature in the winter and increasing it in the summer, taking the train instead of flying for short-distance travel, are relatively simple acts that reduce energy consumption, reduce CO_2 emissions, and in most cases save money.

Energy efficiency has also been a favorite. Despite the fact that it involves capital expenditure, it often (not always) saves money in the long run, especially in regions where energy is relatively expensive. Replacing the old incandescent light bulbs with fluorescent lights and LEDs is an investment that pays off in a few years. Installing an extra layer of thermal insulation under the roofs of houses, always reduces the amount of energy consumed for heating and cooling, but only pays off if energy is not cheap. Modern cars have become more efficient and achieve higher miles per gallon (or kilometers per liter) using aerodynamic shapes, electronic fuel injection systems, and engines with higher compression ratios. Consumer products manufacturers reduce their energy consumption and their CO_2 footprint by reusing waste materials and parts of their process waste heat.

Energy efficiency and energy conservation measures are among the cleanest and least expensive ways to eliminate part of the global energy demand. Instead of generating more energy, energy conservation and efficiency simply satisfy the demand with the consumption of less primary energy and fewer fossil fuels. For this reason, the International Energy Agency (IEA) referred to efficiency as *the first fuel* [8]. However, conservation and higher efficiency do not extend very far in curbing CO_2 emissions. After several decades promoting conservation and higher efficiency in the industrialized nations, most of what can be economically achieved has already been accomplished. More energy efficiency projects may be too expensive to be justified economically and need government subsidies.

In the developing world, lack of capital is an impediment in undertaking energy efficiency projects. Twelve years after the IEA's endorsement of efficiency as the first fuel, global investment in energy efficiency has slowed [9] and energy efficiency projects have been postponed or cancelled in several developing nations [10]. Cheap energy and high cost of borrowing (interest rates) are two of the reasons for the slowing investment in capital-intensive efficiency measures. An example: the average household in Texas consumes 13,128 kWh per year, with approximately 6,000 kWh spent on the air-conditioning (A/C) systems. The A/C consumption may drop by 3,450 kWh (to 2,550 kWh) using Ground Source Heat Pumps (GSHPs), which are available but would cost the owners approximately \$45,000 (after the 30% tax credit). At the 2025 average price of electricity of \$0.142/kWh, homeowners would save \$490/year for an investment of \$45,000, not a sound investment at a meager 1.1% annual rate of return. However, if the electricity price were closer to \$0.45/kWh – a price that is typical in the Northern EU countries – the monetary savings are \$1,380/year, the return is close to 3.5%, and the investment becomes more attractive (in addition, the owners may boast that they are good environmental stewards!). The people in Texas will revolt if their electricity costs jumped from the current \$0.142/kWh to \$0.45/kWh for the purpose of making attractive energy efficiency investments – and many homeowners could not afford the capital outlay, anyway.

In general, high tax credits, lower borrowing interest rates, and higher energy prices favor the adoption of conservation and efficiency measures, but these measures alone are not sufficient to significantly curb the CO_2 emissions. In addition, forced efficiency policies on households are not popular with the majority of the citizens, and any tax credits or other subsidies to industries for efficiency measures must be justified nationally.

3.3 RENEWABLE ENERGY SOURCES

Using renewable energy to satisfy all or part of the global energy demand is a feasible strategy to substantially reduce the global CO_2 emissions. Renewable energy appears in all the roadmaps by international agencies, all the national plans, and all policies to achieve zero-carbon emissions. Since the emissions from electric power plants account for the lion's share of the global CO_2 emissions, the transition of the electricity generation sector to renewable energy has been the first target of policymakers for any future regulations on the reduction of CO_2 emissions. In order to limit the average global temperature rise to a maximum of 2°C by the end of the century, the Intergovernmental Panel on Climate Change (IPCC) recommended that the CO_2 emissions from the electricity generation sector be reduced by 90% or more from the 2010 levels between the years 2040 and 2070 [11]. More recent recommendations and roadmaps by the IEA advanced this goal to the year 2050 and strongly recommended net-zero CO_2 emissions by the global electricity generating industry sooner, a noble but very ambitious goal [12].

While there is a plethora of renewable energy sources (RESs), four of them (solar, wind, hydroelectric, and geothermal energy) account for more than 99.9% of the globally generated electricity by renewables and are expected to be the dominant RESs for the foreseeable future [13]. Of these, hydroelectric and geothermal resources are only available in specific parts of the world. Hydroelectric energy was the first to be harnessed for electricity generation. Most of the important hydroelectric resources in the developed nations were successfully harnessed by the end of the 20th century, leaving only smaller and more costly projects for the future. Parts of the developing world still have undeveloped hydroelectric resources, but these are hardly enough to generate a significant fraction of the electricity needed in the future. Countries like Norway and Nepal may count on hydroelectric energy to supply almost all their electricity demand, but the USA, China, India, and most of the other nations must look for other RESs. Similarly, geothermal resources exist primarily in volcanic areas. Smaller countries such as Iceland, Honduras, and El Salvador may satisfy all their energy needs using geothermal resources, but the rest of the world (including Japan, Italy, China, and the USA that have substantial geothermal resources) cannot realistically rely on this energy source to generate a large fraction of their electric energy [14]. Despite the significant investment and added geothermal power plant capacity in the period 1960–2025, geothermal energy generates between 0.3% to 0.4% of the global electric energy, and less than 0.5% in the USA.

Such considerations leave wind and solar energy, which are available in all the regions of the globe, as the main RESs to supply the future global economy with electricity. However, wind and solar energy suffer from a major disadvantage as electricity generation sources: wind is intermittent, and solar is (at best) periodically variable. The sun shines during the day, and its irradiance may be converted to electricity, but during the night, it cannot supply any electric power. Also, the solar irradiance is higher in the summer days and generates much more electricity in the summer than in the winter. Weather phenomena, such as clouds and rain, also have a substantial impact on the electricity generated by solar energy. Similarly, the wind does not always blow. We have windy days when a great deal of electricity may be generated by wind turbines, but also light breezes and calm days when very little or no electricity may be harnessed from the wind. In short, solar and wind cannot be dependent to reliably generate power when human activities require it, a fact known since the days of Homer, when the 1,186 Greek ships that were scheduled to sail for Troy were stranded in the port of Aulis for lack of favorable winds [15].

3.3.1 Electricity Demand and Supply

The demand for electric power varies widely during the day as well as from season to season. Individuals and families consume a great deal more electricity during the day, when they perform most of their activities, and almost none during the night when they are asleep. Still, the demand for electricity does not vanish during the night, because several industries work continuously, there are lights on streets and buildings, and many electric trains and trolleys continue to operate. More electricity is needed for the operation of the HVAC systems – for cooling in the summers and heating in the winters – than in the springs and autumns, when the weather is milder. The daily and seasonal variabilities of the electricity demand are substantial, especially during the summer days, and in regions that use a great deal of A/C. Typical electricity demand trends on two days, one in the spring and one in the summer, are shown in Figure 3.2, for the ERCOT electricity grid that supplies most of Texas. The

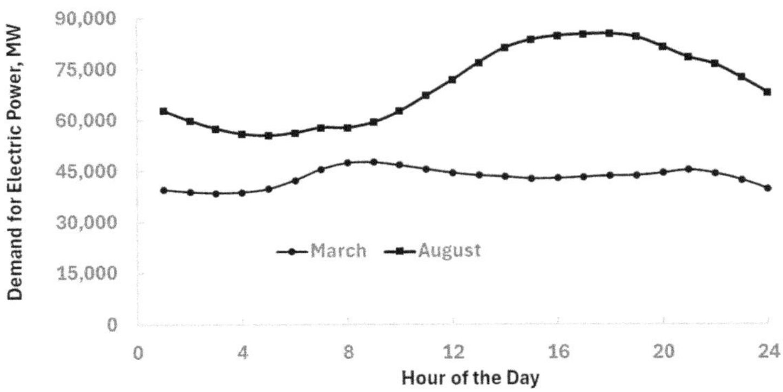

FIGURE 3.2 Typical electricity demand curves on a spring and a summer day in the ERCOT grid.

bump in the demand caused by the A/C usage during the hot hours of the summer day (noon to 8 pm) is apparent in this Figure. One can also observe that the peak demand on the summer day is substantially higher than that on the spring day (by almost 80%), and that, during the summer day, the demand significantly fluctuates (it increases from 55,000 MW to 85,000 MW and then drops to 60,000 MW within a few hours). Since the electricity grid does not store any energy, the electricity generating units must immediately respond to satisfy the fluctuating demand.

The electricity grids transmit the electric power from the generating stations to the consumers but do not store any of the energy they transmit. The power that reaches the consumers is almost simultaneously generated in the power plants. If the demand increases and there is not sufficient capacity to generate power, we have a voltage drop resulting in a "brownout" and if the demand substantially exceeds the supply a "blackout" may occur. During blackout periods, electricity providers cut off the supply in one or more regions of the grid, and consumers do not have electric power. To avoid blackouts, the electricity grid management starts and operates (brings "in line") new power units or increases the generating capacity of existing units. Gas turbines, which largely operate with natural gas and liquid hydrocarbon fuels, have the capability to come in or out of line withing minutes and they are typically used for the daily power demand fluctuations.

A grid that is wholly supplied by the universally available RESs (wind and solar) is incapable of responding to the seasonal and daily fluctuations of power demand. Looking at the power demand of the summer day in Figure 3.2, the demand rises in the late afternoon and remains high in the early evening, at a time when the sun sets and supplies little irradiance to be converted to power. Wind is intermittent, and there may be very little or no wind to make up for the solar power generation drop. Unless the grid has sufficient energy storage capacity (pumped-hydro, batteries, hydrogen, etc.) to generate more electricity and meet the still high demand, a blackout will follow.

3.3.2 The Duck Curve and the Rattlesnake Curve

The term *duck curve* was coined by scientists of the California Independent System Operator (CAISO, the administrative organization of the electricity grid of California) and pertains to a situation when solar energy supplies a great deal of the annual electricity demand [16, 17]. We know that, on a sunny day, the sun supplies higher irradiance to generate higher electric power when it is close to its zenith (at a time close to noon), supplies less power in the early morning and late afternoon, and no power during the night. When more and more solar generating capacity is added to a grid, the power supply increases substantially in the hours close to noon (10 am to 2 pm) when the solar irradiance is at its maximum, but the demand is not always there to pick up the increased supply.

This oversupply situation is shown in Figure 3.3 for the demand in the ERCOT grid during the same summer day, which is depicted in Figure 3.2. The hourly demand is shown in the upper curve of Figure 3.3, and the demand minus the currently (2025) installed wind power supply on the same day is shown in the curve labelled "Demand-Wind." The other curves depict the residual demand (the actual

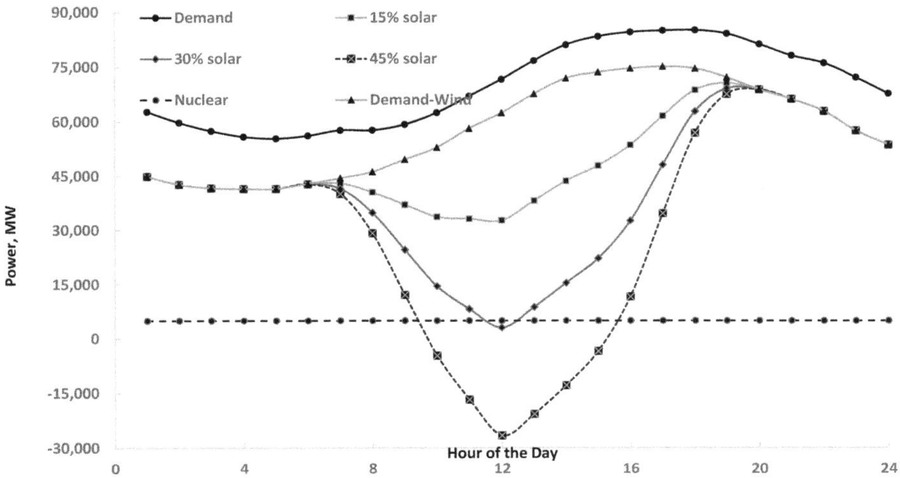

FIGURE 3.3 The *duck curve* indicates the demand-supply mismatch, when the supply of solar energy significantly grows.

demand minus the supply from wind and the additional solar units) for the cases when the solar generating capacity in ERCOT is increased to supply 15%, 30%, and 45% of the total annual electric energy demand. The almost constant power supply of the two nuclear power plants in the ERCOT system is also shown at the bottom of the graph. Since all the solar electricity is generated during the daylight hours, it is observed in Figure 3.3 that, at 45% additional solar energy, the residual demand noticeably drops in the early morning hours and becomes negative in the period 11:00 am to 3:00 pm. During these four hours, there is no sufficient demand in the entire grid to consume the generated electric energy from the solar and wind units. The excess generated electricity must be either stored or dissipated in the grid. Even at 30% additional solar energy, the grid runs into demand-supply problems at noon, because the residual power drops below the capacity of the nuclear units, and the design of most nuclear power plants in the USA does not allow the rapid adjustment of nuclear power to accommodate the increased solar supply.

The situation is similar when the supply of wind energy grows in the absence of large energy storage capacity. Figure 3.4 depicts the demand-supply mismatch for ERCOT during the typical spring day shown in Figure 3.2, when air-conditioning is not needed, and the power demand in the grid is lower and almost flat. The curve labelled "demand" represents the total demand at all hours of the day, and the curve labelled "0%" is the demand minus the power supplied by the already installed wind capacity in December 2025 [18]. The curves labelled "20% and "40%" show the residual demand when the wind generating capacity grows to supply an additional 20% and 40% of the annually generated electric energy. One can see that in the early morning and the late evening hours, when the electricity demand is modest and the wind intensity higher, the residual power demand drops below the nuclear power capacity and becomes negative. During these time periods there is not sufficient

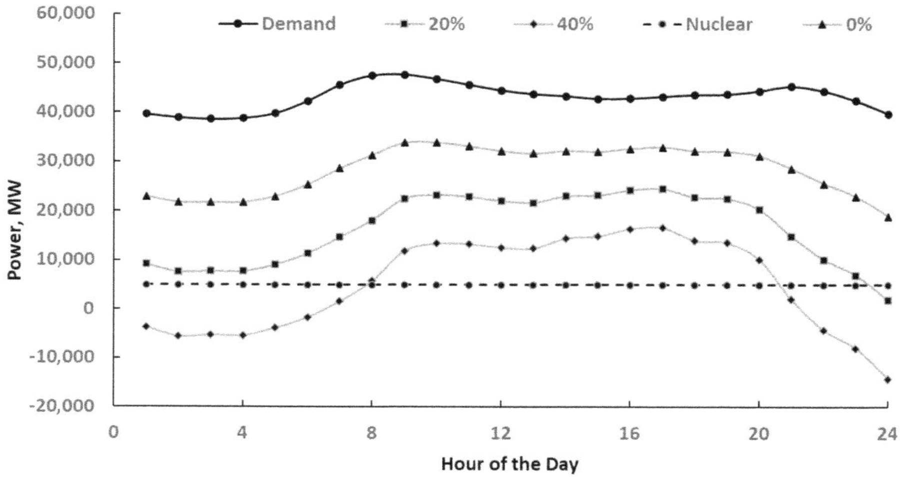

FIGURE 3.4 The *rattlesnake curve* for the residual power demand shows the demand-supply mismatch when the supply of the intermittent wind energy significantly increases.

demand in the entire grid to match the power generated by the intermittent wind. In analogy to the duck curve the types of lines in Figure 3.4 have been given the name *rattlesnake curve*, after the notorious Texan reptile [19].

Another adverse effect of wind intermittency is associated with the prices of other fuels and the cost of electricity, when the wind intensity does not reach its statistically expected level for a long period of time. In an electricity grid where wind is expected to provide a significant percentage of electric power, but the wind intensity is weak over a long period, the electricity generators in the grid must use other energy sources (e.g., natural gas and coal) to generate enough electricity and satisfy the consumers' demand. Between March and September 2021, a series of high-pressure systems over Northern Europe caused a wind drought that resulted in substantially lower wind power generation for the UK, Northern France, the Netherlands, Spain, and Germany. This led to an unexpected and significantly higher reliance on coal and natural gas for electricity generation, at a time when the reserves of these fuels were low. As a result, by October 2021, the electricity wholesale prices in the entire EU increased by more than 200%, while the UK prices skyrocketed by a factor of six [20].

3.3.3 Grid Reliability

Everyone expects that electric power will be available when they switch on their lights and appliances. Frequent blackouts are unacceptable to consumers and can be destructive to industry. Reliability is the most important and desirable characteristic of all electricity grids. However, the most abundant RESs (wind and solar) depend on the weather, and weather is anything but reliable. Photovoltaic (PV) cells do not generate any power in the evenings and generate substantially less power on cloudy days. On calm days the wind turbines are idle.

With the transition of the electricity generation industry to RESs, the vital role of energy storage to the consumers may be summarized with the question: "How do I keep my lights on and operate my A/C unit on a summer evening when the wind is calm?" If the home and city lights are to be on, the stores open in the evenings, and the 24/7 manufacturing plants uninterruptedly operate, it is necessary for an electricity grid to have sufficient power available in order to satisfy the entire consumers' demand. While this is a problem with wind and solar electricity supply, it is not a problem with thermal power plants: coal and nuclear power plants are capable of generating power continuously, and gas turbines may be switched on and off withing minutes to meet demand fluctuations.

One of the fallacies about RESs, often repeated in the popular press, is that a certain wind or solar power unit may provide so much energy (in kWh or MWh units) annually, and that this is enough to supply a number of homes (thousands or millions) with electricity. One can find on websites and the popular press statements such as "...the new wind farm will generate 175.2 GWh annually, enough to power 20,000 homes." Such statements are inaccurate and misleading because they confuse the needed power with annually generated energy [21]. When the wind speed is low and the 20,000 homes need electric power to keep their lights and A/C units on, the new wind farm that generates annually 175.2 GWh is simply unable to power anything. Our daily activities require power, and we need the power immediately, not after a day or a season. When I switch on my living room lights on an evening in July, I require approximately 250 W of electric power for about four hours (for a total of 1 kWh of electric energy). On that evening, I would be very upset if the lights do not go on and the electricity operator sends me a text "...we apologize, our power systems are down because there is no sun and wind energy to be harnessed. We will supply to you the electric energy you require in October, when wind will be plentiful." This is the essence of the electric grid reliability: power must be available when needed, not when the weather is favorable. In a reliable electricity grid the supply must match the demand at any time of the year, not only when the sun shines, and the wind speed is high. Given the daily and seasonal demand fluctuations shown in Figure 3.2, the only way to match demand and supply with RESs is to generate and then store enormous quantities of energy.

Another issue with grid reliability is instabilities with large voltage and frequency oscillations that may lead to failure and regional blackouts – as it happened in Spain during the afternoon of April 28, 2025, and paralyzed the entire country for more than 10 hours. On that day high voltage and frequency oscillations were observed, but the mix of power generators in the Spanish grid did not possess sufficient inertia (from spinning generators/turbines) to absorb them. The result of the grid instability was that at approximately 12:33 p.m., there was a sudden surge in voltage, which the grid was unable to absorb. This led to a cascade of catastrophic events for the grid and, finally, to the blackout of the entire region. The blackout triggered the paralysis of all economic activity, several fatalities, and a cost that is estimated between 2.25 billion and 4 billion euros [22].

Grid operators must constantly balance the electricity supply and demand to keep the voltage and the frequency of the grid within narrow limits and avoid damaging electrical appliances and industrial equipment. Thermal power stations generate electricity with heavy rotating machinery (gas and steam turbines) that have substantial inertia – rooted on the high kinetic energy of the fast-spinning generators. In the case of a sudden loss of power, this inertia is capable of providing a temporary supply of energy that helps maintain the grid's frequency and voltage. Essentially, the inertia of the spinning generators acts as a shock absorber for the entire grid. Solar and wind generators do not inherently possess such inertia, since they typically utilize power electronics to convert their output to the grid frequency, and they cannot react automatically to frequency fluctuations. As the fraction of renewables in the electricity generation mix increases within an electricity grid, the grids' inertia decreases leaving the grid vulnerable to instabilities.

The low inertia vulnerability for a grid can be remedied with the deployment of *synchronous condensers* – large rotating machinery that simulate the spinning turbines and provide inertia without generating active power. Retired conventional turbine-generators can be converted into synchronous condensers. Other methods include the utilization of grid-forming inverters, the integration of fast-reacting energy storage systems (currently batteries and flywheels that can quickly discharge their stored energy and, perhaps, superconducting coils in the future). In addition, synthetic inertia can be included by programming the power electronics to rapidly boost their output while responding to grid frequency drops. Of course, all these modifications to the structure of the grid entail significant investment that must become available to the grid operators. Typically, such investments are borne by the electricity consumers in higher future electricity prices.

3.3.4 Energy Storage

Energy storage systems can store part of the renewable energy when wind and solar units generate an excess of electric power, which the demand cannot absorb. In this case energy stored is not wasted and may be sold later to the consumers. More importantly, energy storage systems supply energy to the consumers, when the RESs generation cannot satisfy all the energy demand. When the sun does not shine and the wind is calm, a storage system takes over the supply side and generates sufficient electric power to satisfy the consumers' demand. The most practical systems that can be used to store energy for electricity generation are:

a. *Pumped-Hydro Storage (PHS)* systems operate like hydroelectric power plants, with water pumped to their reservoirs. When the RESs generate more electricity than is needed, the excess power is diverted to pumps that lift water to an elevated reservoir. When demand exceeds the supply, the same pumps operate as turbines and use the elevated water to generate electricity. The photograph in Figure 3.5 shows a PHS system, the *Alto Tâmega* project, developed by *Iberdrola* in Northern Portugal. This enormous energy storage project has 40 GWh total energy storage capacity and has the capability to supply up to 1,158 MW of power. The reservoir, behind the dam, is supplied by pumps that also double as turbines. In 2025 PHS systems accounted for more than 94% of the total global energy storage capacity [23]. While these systems may store vast amounts of energy, their

FIGURE 3.5 The *Alto Tâmega* PHS project in Northern Portugal has 40 GWh capacity. (Photo courtesy of *Iberdrola Inc.*)

round-trip efficiency (energy delivered divided by energy input) is relatively low, in the range 60–75% [24].

b. *Compressed Air Energy Storage* (CAES) systems utilize underground caverns to store air, which is compressed when the RES electricity supply exceeds the demand. The air can be later used in Gas Turbines (with the addition of heat supplied by natural gas). The round-trip efficiency of CAES systems is also low, in the range 45–60%.
c. *Batteries,* both solid-state and flow batteries, have significantly higher round-trip efficiency, that may reach 90%, depending on how fast they are charged and discharged – with fast charging always dissipating more energy. Although most batteries are made of relatively expensive materials, several electricity generation corporations are currently developing battery storage systems for short-term energy storage, to be used for *price arbitrage,* that is, to store electric energy when its price is low and sell it when the price increases. The *Flow Batteries,* which primarily use vanadium ion electrolytes, may store and deliver high quantities of energy, but their round-trip efficiency is relatively low – currently at about 75%.
d. *Hydrogen* storage is promising because this gas is stable and may be stored for a long time for seasonal energy storage (e.g., using the high winds in the spring to supply the A/C units in the summer). Large quantities of hydrogen may be produced by water electrolysis. If necessary, hydrogen may be transported over short distances (less than 300 km) with relatively low transportation cost. Hydrogen is converted back to electricity in fuel cells. The round-trip efficiency of hydrogen storage systems is in the range of 50–75%, with the higher values corresponding to solid oxide fuel cells [24].

FIGURE 3.6 The chilled water storage building in the DFW airport, with the third terminal of the airport in the background. (Courtesy Mr. Rusty T. Hodapp.)

e. *Thermal storage* makes use of chilled water to provide the cooling effect of A/C in the summer, or hot water to heat buildings in the winter. The photograph in Figure 3.6 shows the reservoir building that contains a well-insulated tank for the chilled water used for A/C supply in the Dallas-Fort Worth airport (DFW, with cars and airplanes shown for scale comparison). The chilled water circulates in the five terminals of the airport via a network of large subterranean pipes during the peak electric power demand hours of the summer and keeps the interior temperatures of the airport terminals at the desired level. The water is chilled during the nights, when electricity is cheaper in the region, and the supply of "coolness" to the terminals occurs in the early afternoon, when electricity is more expensive.

f. *Other energy storage* systems that are currently used or are in the development stages are flywheels, gravity systems, supercapacitors, superconducting coils, and hybrid systems that consist of two or more of the other systems. Most of them serve particular needs (e.g., flywheels are capable of generating very high power for short time to produce plasma or correct for frequency fluctuations). The contribution of these systems to the global energy storage in 2025 was less than 1%.

It must be noted that fossil and nuclear fuels are actually "stored energy" that may be used at any time we need it. Nuclear elements were formed at the birth of the planets, and fossil fuels were formed several millennia ago. They have been stored in the earth's interior for centuries and are available to be mined and used at will.

3.3.5 Power Operating Factors – the Needed Power for Transition

The power operating factor (POF) – sometimes called capacity factor (CF) – is an important characteristic of all electricity generation units. It is defined as the total energy annually generated by the unit divided by the nominal power it may generate and the number 8,760. The latter is the number of hours in a (non-leap) year. For example, if a nuclear power plant has a nominal power-generating capacity of 2,000 MW and operates for fifty weeks in a year (8,400 hours, leaving two weeks for maintenance, and generating 16,800,000 MWh annually), its POF is $16{,}800{,}000/(2{,}000 \times 8{,}760) = 0.959$ (or 95.9%). Nuclear and fossil fuel power plants may operate continuously (at the will of their operators) and their POFs can be close to 100%.

Because the sun does not always shine and wind is not always available, the POFs of the solar and wind units are significantly lower. Figure 3.7 depicts the hourly electric energy generated by all the wind turbines in the ERCOT grid during the year 2023. The obviously high variability of the hourly data – from a high of almost 26,300 MWh to a low of 38 MWh – reflects the weather and wind velocity fluctuations in the ERCOT region. For the years 2020–2025, the POF of the wind units in ERCOT was 32.7%. PV solar units exhibited even lower POFs, close to 21%.

The high discrepancy among the POFs of thermal and renewable energy units indicates a significant weakness of solar- and wind-generated power: in order to substitute fossil fuel units and generate the annual quantity of electric energy demand, significantly more renewable power needs to be installed than the power of the substituted thermal units. For example, if the 2,000 MW nuclear power plant with POF = 95.9% is to be substituted with renewable power units, the installation of $2{,}000 \times (95.9/32.7) = 5{,}865$ MW of wind power or $2{,}000 \times (95.9/21) = 9{,}133$ MW of PV (or a combination of the two) needs to be installed. Energy storage will also be needed to meet the fluctuating demand and supply. In the latter case, even more

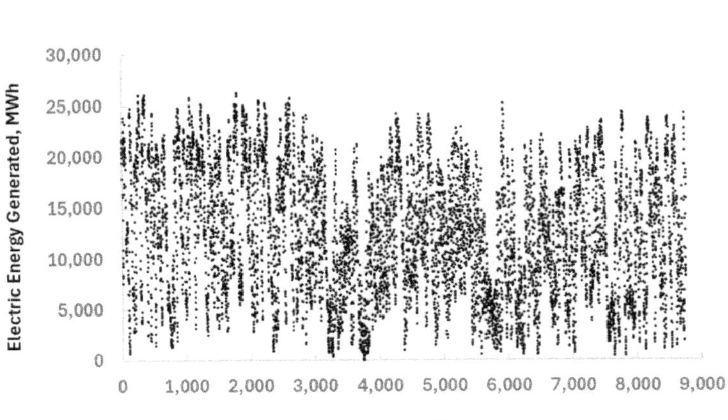

FIGURE 3.7 The electric energy generated by wind in the ERCOT grid during all hours of 2023.

installed power is needed to account for the energy losses in the energy storage-regeneration processes, which are necessary for reliability. Simply put, the installed energy capacity of the renewable units must be significantly higher than the power capacity they substitute, signaling high investment expenses. Of course, this is mitigated by the fact that the sun and wind energy are free and that solar and wind units do not have any fuel expenses.

A case study that illustrates the importance of POFs in the transition to RESs is that of Germany [25]. In the year 2000, Germany had an installed power capacity of 121 GW (1 GW = 10^9 W or 1,000 MW) and generated 577 TWh of electricity – 60.1% from fossil fuels, 6.6% from renewables and 33.3% from other sources, primarily nuclear. After substantial investments in wind and solar energy during the first two decades of the 21st century, in 2019, Germany generated 607 TWh of electric energy (5% increase) with an installed power capacity of 218.1 GW (80% increase). The renewables share in the electricity generation jumped to 41.1% by 2019, while the fossil fuels share decreased to 47%. The significant and necessary rise of the installed power capacity, while the total electric energy generated is almost flat, is explained by the low POFs of the renewable power units: the average POF of the solar units in Germany was a mere 10%, (which is not surprising since most of Germany, with the exception of Bavaria, is perennially cloudy) and the average POF of all renewable units was close to 20% [25]. The older fossil fuel and nuclear units operated at an average POF of 85%. In its path to electricity generation transition, Germany also discovered that it needs to keep the older fossil and nuclear power plants operational to meet the demand during cloudy and calm days.

3.4 MORE NUCLEAR ENERGY

Electricity generation in nuclear power plants does not entail any CO_2 emissions, a fact that makes this option attractive. The technology of electricity generation using nuclear reactors is well-developed, with numerous OECD countries having more than seventy years' experience with nuclear power plants. A few developing countries have joined the ranks of nuclear reactor operators since the 1990s, but most of the developing world does not have nuclear know-how. Given the dangers of a large-scale nuclear accident, lack of nuclear reactor technology and experience, geopolitical instability, and lack of enforceable international safety standards, this option is a rather risky solution for most developing nations. A concerted international program of nuclear technology transfer to developing nations – or the construction and operation of nuclear power plants by international organizations and consortia – as well as international oversight of the nuclear reactor design and operations, on a global scale, can become a viable long-term solution for the reduction of GHG emissions using more nuclear energy.

Nuclear reactors are capable of supplying electricity at all times of the year. In principle, a global network of nuclear reactors (with some storage capacity) is capable of supplying the entire global electricity demand, and there are several scientists that promote such a solution, without the addition of a great deal of solar and wind energy [26]. However, before this solution is widely adopted, and before nuclear energy becomes suitable as a viable solution to the GCC problem, the global community will have to address two problems:

1. The safety concerns of the general population, which were fueled by the accidents at Windscale (1957), Three Mile Island (1979), Chernobyl (1986), and Fukushima (2011). Safety concerns of individuals – often summarized in the slogan "not in my back yard" (NIMBY) – are responsible for the decommissioning of a large number of nuclear reactors in the OECD countries since the 2010s.
2. Environmental and ecological problems, most notably the long-term storage of nuclear waste. Nuclear waste needs to be stored for millennia, and most countries do not have permanent repositories for it.

The NIMBY effect is particularly evident in the 21st century within several OECD countries, which have stopped building new nuclear power plants or have decided to shut down their existing nuclear reactors. Figure 3.8 demonstrates this effect for the USA [27]. The Figure depicts the annual electric energy generated by the nuclear power plants in this country (left axis) and the percentage of the total annual electric energy that was generated by the nuclear reactors (right axis). Figure 3.8 shows that, after the mid-1990s, the total electric energy generated by nuclear reactors has stagnated at about 780 TWh. The percentage of electricity generated by the nuclear power plants in the USA has actually decreased from its peak of 20.6% in 2001 to approximately 18.2% in 2025.

Unlike the other widely available non-carbon electricity power units (wind and solar) nuclear power plants are capable of operating 24/7 and in all seasons. Their POFs are around 90% and may be even higher. This means that significantly less generating power capacity (in MW) is needed when nuclear units substitute the fossil fuel units, than when wind and solar units are used in such substitutions. A 2020

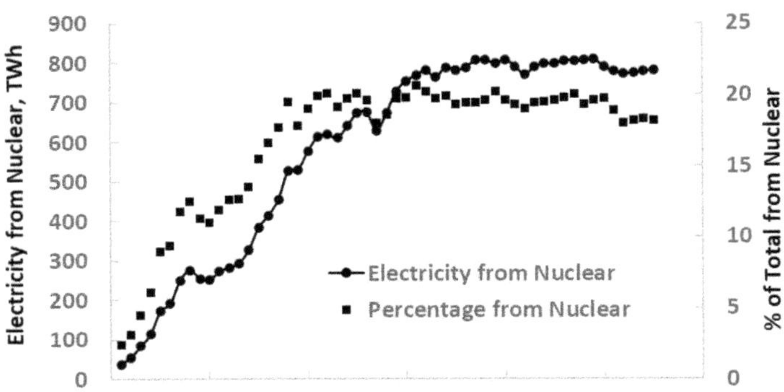

FIGURE 3.8 The "not in my back yard" effect on nuclear-generated electricity in the USA. Total electricity generated by nuclear reactors (left axis) and percentage of total electricity generated by nuclear energy (right axis).

study on the decarbonization of the ERCOT system (which currently has approximately 5,000 MW nuclear capacity) concluded that an additional 36,700 MW nuclear capacity (accompanied with 35 TWh storage capacity) would be sufficient to satisfy the electricity demand and generate all the electric energy needed in the system [28]. For comparison, for the same substitution with solar PV units, the additionally needed power capacity is 185,000 MW, and for substitution with wind it is close to 90,000 MW. The needed storage capacity is about the same at 35–40 TWh.

The past stagnation of the global nuclear generating capacity notwithstanding, in early 2026 there are a few bright spots about the future contributions of nuclear energy. After the significant expansion of wind and solar units in the first three decades of the 21st century, the political establishment has realized that renewable energy alone will not lead to decarbonization and that nuclear energy (despite its disapproval by the public that insist on NIMBY) must be recruited to continuously generate electric power. New nuclear power plant construction is seeing a resurgence in numerous Asian countries. Several OECD nations that were planning to shut down their nuclear reactors are changing course. In the USA, the New England States are calling for more diverse energy strategies with combinations of nuclear and renewables – a significant shift from the rhetoric of the early 2000s. However, a problem for the immediate deployment of more nuclear power plants in the USA and other countries is litigation with communities and environmental groups, which results in long periods of time for the planning and completion of nuclear projects: it took 18 years for the two Vogtle nuclear reactors (in Georgia, USA) from planning to commercial operation. Planning started in 2006, construction was approved in 2009, construction began in 2013, the first unit became operational in 2023, and the second unit became operational in 2024. This is a long time to wait for additional zero-carbon electricity!

3.4.1 Small Modular Reactors

The vast majority of the electricity-generating nuclear reactors have high generating power capacity, typically 1,000 MW or higher for each reactor. Nuclear reactors are closed systems that keep all their fuel and nuclear waste (the uranium fission products) for a period of 18–24 months. After that time, the fuel cycle is repeated, the reactors are refueled, and remain shut for another 18–24 months. Because of this long refueling period, nuclear reactors contain large amounts of radionuclides, which pose a great risk to the environment if they are released. There are hundreds of smaller nuclear reactors that are constructed to power nuclear naval vessels (chiefly nuclear submarines and air carriers) as well as those that serve as neutron sources. These are the Small Modular Reactors (SMRs), that are much smaller than the commercial reactors, contain lesser quantities of radionuclides, and may be used to supply smaller quantities of electric power. The engineering community, which has designed and operates the naval reactors, has the know-how and significant experience in the design and operation of SMRs.

The use of SMRs, which do not release any CO_2, may supplement wind and solar power in zero-emissions communities in the future and significantly assist with the decarbonization of the electricity generation industry. It is characteristic of the importance of nuclear energy in mitigating the GCC that, in the third decade of the

TABLE 3.1
Wind or Solar Power Capacity, Renewable Power Reductions and Substitution Rates with the Addition of a 20 MW and a 40 MW SMR to Achieve Zero Carbon Emissions in a Community of 50,000 Homes

Nuclear Power, MW	Wind Power, MW	PV Power, MW	Wind Reduction, %	Solar Reduction, %	Substitution Rate, Wind, MW per MW Nuclear	Substitution Rate, Solar, MW per MW Nuclear
0	243	552				
20	185	411	23.9	25.5	2.90	7.05
40	129	283	46.9	48.7	2.85	6.73

21st century, the roadmaps for zero-carbon as well as the future energy plans of most developed nations include large or smaller nuclear reactors to assist the decarbonization of the electricity generation industries [29, 30]. A computational study investigated a zero-carbon community of 50,000 homes in North Texas, powered with SMRs, wind, and solar units [31]. Table 3.1 shows the results of this study when SMRs of 20 and 40 MW are developed to supply some of the needed power in the community. The addition of nuclear generating capacity significantly reduces the required wind and solar power. And, because the POFs of nuclear reactors are close to 100%, 1 MW of nuclear power substitutes for approximately 3 MW wind capacity and close to 6 MW solar generating capacity.

While the SMRs have not yet contributed a great deal of energy, in early 2026, there are several of these reactors in the construction stage throughout the world, including two in the USA – one in Wyoming with 345 MW capacity and the second a 50 MW in Tennessee. Both SMRs will provide power support to data centers.

3.4.2 Problems with Nuclear Reactors

Nuclear energy has always evoked trepidation in the general public. The introduction of nuclear energy to the world was one of utter destruction following the bombings of Hiroshima and Nagasaki at the end of World War II. This perception was exacerbated by the fear of nuclear war during the Cold War period, 1945–1989. The general public has always been apprehensive and mistrustful of nuclear fission, even when it is applied for peaceful purposes.

The highly publicized accidents in nuclear power plants did not help allay this public perception. The bad publicity of the Three-Mile Island nuclear power plant accident, and the environmental disasters following the accidents in Chernobyl and Fukushima have created a cloud of mistrust in the general public about nuclear reactors and nuclear energy, in general. This public sentiment about nuclear reactors (often summarized by the NIMBY expression) has immensely contributed to the slow-down of nuclear industry worldwide. While several older reactors have been decommissioned, only two new nuclear reactors since 2000 were constructed in the USA; Italy, Germany, Taiwan, Lithuania, and Kazakhstan have shut down their nuclear reactors; and Spain with Switzerland are planning to decommission their own reactors by 2035. China, India, South Korea, and the UK are among the countries where new nuclear power plants are under construction, and a few more countries are considering nuclear energy for the future.

The notoriety associated by the general public with nuclear reactors has created a situation that any nuclear reactor incident – which has the appearance, perception, or portrayal as a serious accident by the information media – may cause the cancellation of entire national nuclear programs (as happened in the USA in the period 1990–2020). Since reactor safety is of paramount importance in the nuclear industry and since modern reactors are built with high safety standards, the nuclear industry must convince the general public that their reactors do not only contribute to generate carbon-free electricity but are also safe to operate and will not be the cause of Chernobyl-kind disasters.

The permanent disposal of nuclear waste is a second issue that needs to be addressed. All nuclear reactors generate low-level radioactive waste that must be sequestered and stored for centuries. However, NIMBY is still active in this area, with states and regions refusing to accept nuclear waste to be settled within their boundaries. For lack of a permanent nuclear waste storage facility, the nuclear waste generated in the USA is currently stored in pools of water close to the reactors that produced it. The construction of a permanent Federal repository facility in Yucca Mountain, Nevada, was first voted for and funded by Congress in 2002 and then nixed by the Department of Energy in 2009. European nations (France, Belgium, Germany) have temporary central storage facilities, but only the UK has a permanent nuclear waste storage facility – in Sellafield (formerly Windscale), the site of the 1957 power plant accident. P.R. China has also established a permanent underground nuclear waste repository – in Longhe within the Gobi Desert.

It must be noted that the scientific community has offered a viable solution for nuclear waste: it may be safely processed and stored in underground caverns and subterranean tunnels. The processing of nuclear waste includes densification and vitrification (glassification) to reduce the volume of the waste and to maintain its cohesiveness. The final storage of the processed waste is in steel containers that are placed underground in dry and stable geological formations. The legal stewardship of the stored nuclear waste is still a sticky point. Instead of leaving stewardship to an ephemeral corporation that may declare bankruptcy in the future, stewardship of the waste must be borne by national governments, the only entities that can guarantee the continuous and safe operation of the sites as well as the maintenance of the storage facilities over long periods of time.

3.5 SUBSTITUTION OF COAL BY NATURAL GAS

This can be a short-term solution that would significantly reduce the global CO_2 emissions. Both coal and natural gas power plants convert their fuels to heat and use the heat to generate electric power. Carbon, the main constituent of coal, produces 0.11 kg of CO_2 to deliver 1 MJ of heat in the steam boilers of the coal power plants. Methane, the main constituent of natural gas, produces only 0.05 kg of CO_2 to deliver 1 MJ of heat (for those who prefer the British units, 1 MJ is approximately

1,000 Btu). In addition, the modern combined cycle natural gas units exhibit thermal efficiencies close to 60%, while the average coal power plant has 33% thermal efficiency [32]. Simply put, 1 MJ of heat goes much further to generate electricity with modern natural gas units. Calculations show that, if all the coal power plants in the USA were to be replaced by the more efficient combined cycle natural gas units, the annual CO_2 emissions of the USA electricity sector would be reduced by approximately 611 million tons, or 12.7% of the total USA emissions. When the same calculations are applied to the globally generated 10,500 TWh of electricity from coal power plants, the global CO_2 emissions drop by 9,450 million tons, a 24% reduction of the global emissions.

Without doubt, the conversion of the coal power plants to gas-fired plants would have a meaningful effect on the reduction of the global CO_2 emissions. However, there is a caveat for this solution: most countries that burn coal for their electricity (including the highly populated China, India, the EU, Pakistan, and Indonesia) do not have sufficient natural gas reserves to satisfy this substitution. The importation of the vast quantities of natural gas, needed to replace the domestic coal, requires additional infrastructure (pipelines, liquified natural gas terminals, etc.). The increased demand for natural gas will also increase prices. The combination of the two effects may cause the substantial increase in the cost of electricity from gas-powered turbines.

3.6 NUCLEAR FUSION ENERGY

Nuclear fusion is the source of energy of the stars including our own sun. Fusion occurs when the nuclei of light isotopes, typically hydrogen isotopes, combine to form the heavier helium nuclei. Calculations show that when four kg of hydrogen fuse, they produce a little less than 4 kg of helium and release 257×10^{13} J, an extremely high amount of energy. The conversion of approximately 180 tons of hydrogen to helium would generate sufficient energy to satisfy the energy demand of USA for an entire year [14]. A few tons per second of hydrogen converted to helium provide all the energy our sun broadcasts to the entire universe. Nuclear fusion is very powerful!

The very first demonstration of man-made fusion, was the detonation of the hydrogen bomb (the H-bomb), which released a tremendous amount of energy. If nuclear fusion reactions were to happen in a controlled manner, the energy released could be used for the peaceful production of electricity. Given that the earth has almost unlimited quantities of hydrogen (in the water of the oceans) controlled nuclear fusion reactors would solve the energy problem of humanity for billions of years. However, controlled fusion on a scale that would produce significant power has not been achieved yet. Several technological breakthroughs are still needed to harness the hydrogen fusion energy for peaceful applications.

There has been some progress toward achieving fusion-generated electricity. A fusion "ignition experiment" conducted in the Lawrence Livermore National Laboratory (in December 2022) lasted for a few milliseconds and produced more heat than the coherent light supplied by the laser beams of the Laboratory. However, other than ignition for a few milliseconds this approach may not be used for continuous electricity generation. The International Thermonuclear Experimental Reactor (ITER) is the outcome of international collaboration (EU, China, India, Japan,

Russia, South Korea, and the USA) and is constructing the world's biggest nuclear fusion reactor in southern France. Its goal is to demonstrate controlled fusion for at least 400 seconds and generate 500 MW of electric power. However, this international project has been plagued by cost overruns and delays, with the first production of plasma delayed from 2025 to 2034 and the actual operation starting in 2039 [33]. In summary, the fusion electricity generation technology may (or may not) become a reality toward the end of this century and is not expected to contribute much to curb the CO_2 emissions by 2050 or even by 2070.

3.7 CARBON CAPTURE AND SEQUESTRATION

It is technologically feasible to capture the produced CO_2, transport it to a suitable location, and sequester it. The technology of CO_2 separation from the other combustion gases (excess oxygen, nitrogen, etc.) and its capture involves the use of membranes, chemical separation (e.g., with amines or carbonates), or liquefaction. Removing CO_2 from the air (where its concentration is very low at approximately 429 ppm or 0.043%) requires 3.5 times more power than removing it from the exhausts of combustion engines. Since it does not make any sense to try and capture CO_2 at moving sources (e.g., the exhausts of cars, trucks, ships, airplanes, etc.) the optimum places for CO_2 removal are stationary production sites (coal power plants, gas turbines, cement factories, etc.) where its concentration is close to 15% and the power required for its separation from the other exhaust gases is substantially lower [34].

Once the CO_2 is separated from the other exhaust gases, it must be pressurized to be transported to the storage location. Optimum transportation at ambient temperatures requires supercritical pressures, typically 100–120 atm, and consumes substantial amounts of power. For a coal power plant, the power required for the separation and pressurization of CO_2 amounts to 35–40% of the total power generated by the plant. For a cement production unit (where part of the waste heat may be used) the required power is less but still high (0.802 MJ per kg of CO_2 or 0.674 MJ per kg of cement) [35].

3.7.1 Inland Sequestration

Depleted oil fields, underground coal deposits, and underground aquifers have been proposed for the long-term sequestration of CO_2. However, one has to keep in mind the following facts about the underground CO_2 storage:

1. There is not enough empty space in large geological formations, such as depleted oil and gas fields, to hold the vast quantities of CO_2 produced daily. The total annual global crude oil production is approximately 36 billion barrels, while the total generated CO_2 (pressurized at 120 atm and stored underground at 60°C) is 560 billion barrels. At the very best, less than 7% of the produced CO_2 may be stored in depleted oil fields. In addition, the gas has to be transported from the power plants to the oil fields at considerable energy expense. The entire storage capacity of the famous Uchira formation (offshore, in the west of Norway) could only store one to two years of the anthropogenic CO_2 emissions.

2. Most of the proposed underground formations are in depths where the temperatures are relatively high, 50–80°C. Since supercritical CO_2 is one of the most powerful solvents, at these high temperatures and pressures, it is very likely that this chemical will find its way out of the reservoirs and will escape in the atmosphere, instead of being stored for a few centuries. It would be an environmental disaster if the highly pressurized CO_2 that is stored underground in 2026 finds its way to the earth's surface and the atmosphere in 2036.
3. Storage in underground aquifers makes the water acidic (and non-drinkable). In addition, the gas causes the dissolution of carbonates and minerals in the aquifer, weakens the surrounding rocks, and creates pathways for CO_2 and brine leakage in the atmosphere [36].
4. The capital cost for the development of the CCS infrastructure (CO_2 separators, compression systems, pipeline systems, injectors, etc.) is very high. The amortization of this cost and the energy cost for CO_2 capture and transportation will increase the cost of electricity generated by a factor of 3 to 4 [34].

All in all, the inland sequestration of CO_2 (in the very limited space available) is very costly, consumes a great deal of power, needs long pipelines, and does not provide any assurance that the gas will remain underground for at least 200 years.

3.7.2 Deep Ocean Sequestration

Deep ocean storage of CO_2 is achieved by injecting it, at high pressure, either by ships or by permanent pipelines. There is an interesting fact about the ocean storage of CO_2: because the molecules of CO_2 are planar and angular, the dissolution of the gas in fresh or saline water increases the density of the aqueous solution and makes it slightly heavier than the surrounding saline water [37]. The heavier solution of CO_2 and saline water naturally sinks in the ocean and does not rise toward the surface, where the gas may escape into the atmosphere. This property makes the ocean sequestration of CO_2 significantly more risk-free than storage in geological formations. Detailed studies show that if CO_2 is injected as small bubbles in the ocean at depths more than 600 m, it will dissolve in the saline water, and the solution will sink to the bottom [34].

With an average depth of 3,800 m, the oceans cover more than 70% of the earth's surface. The capacity of the ocean to absorb and store CO_2 is in the order of 10^{19} tons of the gas, much higher than the annual emissions [38]. At the current emissions rate the deep ocean has the capacity to sequester the equivalent of 300 million years of CO_2 produced by anthropogenic activities. However, as in the case of land sequestration, this does not reduce the heavy energy expenditures for the capture, transportation, and pumping of the gas at the appropriate depths of more than 600 m. These processes are energy intensive, and the needed infrastructure – separation facilities, compression systems, pipelines, injection, monitoring, etc. – is very costly [34]. A 2008 case study by *Greenpeace International* [39] estimated that the cost of CO_2 sequestration would be as high as 92 US$ per ton ($144 in 2025 US$). The same study concluded that spending such sums for carbon sequestration "...diverts resources away from real solutions."

Apart from the energy consumption and high costs, there are environmental considerations with the sequestration of this gas in the oceans: the injected CO_2 in the seawater reduces the pH of the oceans and may pose a threat to aquatic life. Several international treaties including the *London Convention on Ocean Dumping* – the USA is a signatory of this treaty and also enacted its own *Marine Protection, Research and Sanctuaries Act (MPRSA)* – prohibit the storage of industrial waste in the deep ocean. Since CO_2 emissions from power plants are considered industrial waste, such treaties apply to the deep ocean CO_2 storage.

In summary, the capture, pressurization, and transport of CO_2 require substantial amounts of energy, and the needed infrastructure is costly. The sequestration of this gas in underground formations is risky, because the gas may escape into the atmosphere after a short time. Deep ocean sequestration is less risky but entails significant environmental and ecological problems.

3.8 REFORESTATION AND BIOFUELS

Since plants absorb CO_2 from the atmosphere and store it as carbohydrates, the planting of trees on unused land (without burning the produced wood as biomass) has been suggested as a viable CO_2 storage method. While reforestation has other beneficial effects for the regional and global environment (restoring wildlife habitats, increasing biodiversity, preventing soil erosion, and helping to regulate water cycles) its impact on the atmospheric CO_2 concentration is very weak, because of the large magnitude of the daily anthropogenic emissions. For example, a typical 400 MW coal power plant produces daily 111,000 tons of CO_2. It will take approximately 9,500 fully grown pine trees to remove this daily amount of CO_2, or almost 3.5 million pine trees per year! If reforestation is employed on a global scale to sequester this gas, we will need to plant an area larger than that of Germany or Finland every year with trees (which will take several years to grow and fulfil their sequestration potential). Given the global scarcity of arable land and the need to produce food, reforestation is not a viable option for CO_2 sequestration and GCC mitigation. The increasing population of the planet needs the available fertile land for food production, not carbon storage.

The conversion of biomass to produce biofuels has also been touted as a way to reduce global CO_2 emissions. Burning biomass or its products is neutral to the atmospheric accumulation of CO_2 because the plants have absorbed the gas from the atmosphere to store it as biomass. The production of ethanol fuel (derived from sugar cane, switchgrass, and corn) to be used as liquid fuel is a well-known technology widely used in Brazil, USA, the EU, China, and several other countries. However, the energy required for ploughing, fertilization, harvesting, and converting the crop biomass into biofuels is substantial. In the case of ethanol from corn, the consumed energy can be higher than the energy derived from the produced ethanol. Table 3.2 summarizes the principal results of four studies [40–41] on the total energy consumed for the production of corn from 1 hectare of land (2.47 acres) and the energy output-to-input from the production of ethanol from the harvested corn. Two of the

TABLE 3.2
Total Energy Inputs, in GJ/ha, and Energy Output-To-Input Ratio for Corn to Ethanol Conversion

Study	Total Preprocessing	Total Processing	Total Energy Consumption	Energy Out / Energy In
Pimentel, 2003	33.8	51.4	85.2	0.87
Patzek, 2004	29.2	51.4	80.6	0.92
Shapouri et al. 2002	22.3	46.7	69	1.07
Wang et al., 1997	19.8	45.9	65.7	1.13
Average	26.28	48.85	75.13	1.00
% standard deviation	24	6	12	12

studies show a net energy deficit [40, 41], and the other two, which did not consider the wear (and need for replacement) of agricultural equipment, show a very small and inconsequential energy surplus [42, 43].

Despite the very low energy yield, the USA produces more than 16 billion gallons of ethanol annually, to be blended with gasoline. About 12% of this quantity is exported – chiefly to Canada, the UK, and India [42]. The ethanol-for-fuel production is heavily subsidized, at more than $1 per gallon. While the subsidies assist the communities of farmers, who need national support, the energy benefits of corn-to-ethanol production are highly questionable.

The use of agricultural waste (corn stover, tree clippings, bark of harvested trees, etc.) as a fuel in power plants does not entail this energy deficit disadvantage (since the plant/tree growth and harvesting serve another purpose). However, the biomass/waste still needs energy for transportation – a formidable and energy-consuming task because of the large volumes of waste that need to be transported. For example, the 3,900 MW Drax power plant in the UK has the capacity to burn 32,000 tons of wood pellets daily, two-thirds of which are shipped from Louisiana and Mississippi in the USA. The pellets' trip is more than 5,500 miles and consumes a great deal of energy in loading-unloading and transportation. Another problem with wood burning for electricity generation is that unscrupulous market players have abused the wood pellet supply by chopping down healthy trees to sell as agricultural waste, as in the case of the old and protected forests in the Gurghiu mountains (Romania) that were harvested to produce wood pellets, which were exported to be burned as waste in Denmark [43].

As in the case of reforestation, growing biomass for energy is not a viable global option to reduce CO_2 emissions and the available arable land is better to be used to grow food for an increasing global population.

3.9 OTHER MITIGATION SCHEMES

3.9.1 Commercial Uses of Carbon Dioxide

Fertilizers, carbonated drinks, dry ice, and fluids used for enhanced petroleum recovery contain CO_2. It has often been proposed that more applications are found for high quantities of CO_2 to be captured at the exhaust of power plants and then sold commercially. However, the global market for CO_2 is close to 230 million tons, more than half of which (130 million tons) is used for the production of urea for fertilizers [44]. The 230 million tons of total CO_2 demand represents only 0.6% of the global emissions (39.1 billion tons) an almost negligible fraction. In addition, the commercial use of CO_2 does not necessarily mean that this greenhouse gas is actually sequestered for a long time. The urea in fertilizers is hydrolyzed in the soil to generate nitrogen compounds and CO_2, which diffuses through the soil and enters the atmosphere. Most of the CO_2 in enhanced oil recovery applications finds its way back to the atmosphere; and everyone knows that the CO_2 in carbonated drinks is not stored in the human body, but it is eructated in a short time. Close to 85% of the commercially used CO_2 (as spent fertilizers, in enhanced oil recovery, carbonated drinks, etc.) is not sequestered but recycled in the atmosphere within a year. All in all, the global "CO_2 market" does not have the capability to absorb and sequester a meaningful fraction of the huge amounts of anthropogenic CO_2 emissions.

3.9.2 Regenerative Agriculture

Regenerative agriculture entails a set of farming practices aiming to improve the soil, rather than convert it to a carbon sink. Plant photosynthesis removes a fraction of the CO_2 from the atmosphere. Instead of using the non-edible crop residue and burning it as waste or biomass, regenerative practices bury it in the soil and use it to enhance the carbon content of the soil. Thus, part of the atmospheric CO_2 is eventually sequestered into soils with high organic content [45]. Widespread regenerative agriculture processes are beneficial for soil health, but they have extremely limited potential to sequester meaningful quantities of CO_2, because a large part of the global crop residue is used as animal feed, and the supply of the remaining residues as carbon sinks in the soil is limited [46].

A caveat in the extensive use of soil regeneration and buried organic biomass for carbon sequestration, is that if the organics in the wet soil undergo anaerobic decomposition, they produce methane, a GHG more potent than CO_2, which easily diffuses through the soil and enters the atmosphere to exacerbate the GCC problem.

3.9.3 Seeding the Oceans

The oceans are a large sink of CO_2 and, via natural processes absorb approximately 30% of the anthropogenic emissions. Iron- and nitrogen-rich ocean fertilizers promote the rate of CO_2 absorption by aquatic organisms, which form more complex organic compounds in the oceans. However, the same fertilizers contribute to the eutrophication of the aquatic environment, a process that causes oxygen depletion and the demise of fish. Although this option has not been tried on a large scale, it is doubtful that it will result in a significant and meaningful atmospheric CO_2 reduction. In addition, the chemical alteration of the aquatic environment entails the risk of widespread ecological changes as the aquatic species are transformed from a food source for humans to CO_2 sinks.

3.9.4 Seeding the Atmosphere with Particles

Small particles and aerosols in the upper atmosphere reflect part of the sunlight and allow less irradiance to reach the biosphere. It has been suggested that we seed the upper atmosphere with highly reflective micro- and nano-size particles to reduce the solar irradiance and essentially "cool" the planet. Because eventually the particles settle, atmospheric seeding will have to be repeated periodically – likely every two years. This and similar approaches to GCC mitigation have only been experienced with computer models, without any attempt to independently validate and verify the models and computer codes. Such solutions to the GCC problem do not embody the cause-and-effect relationships of GCC but attempt to cool the planet almost uniformly, addressing the effect rather than the cause. The scheme to seed the upper atmosphere with particles is poorly understood and enormously risky since it is based only on unverified computer models. Apart from the fact that it is costly, seeding periodically the atmosphere with particles entails the risks of causing severe regional climate changes with catastrophic local consequences, human health risks, and risks to the global agriculture.

Because of their unpredictable high risks, seeding the oceans and the atmosphere (processes often referred to as *geoengineering*) are not considered as options for the future, and such schemes are excluded from recommended "pathways" and "roadmaps" for GCC mitigation strategies by international organizations and national governments. The IPCC-2019 report has been specific on schemes of seeding the atmosphere and the oceans: "...although some [seeding] measures may be theoretically effective in reducing an overshoot, they face large uncertainties and knowledge gaps...as well as substantial risks and institutional and social constraints to deployment related to governance, ethics, and impacts on sustainable development. They also do not mitigate ocean acidification" [47].

3.9.5 Conversion to Minerals

Several metallic minerals exposed to carbonic or bicarbonic acid (essentially aqueous CO_2) will form carbonate or bicarbonate salts that are solids and precipitate out of the aquatic solution. For example, magnesia (MgO) converted to magnesium hydroxide [$Mg(OH)_2$] will combine with CO_2 to form magnesium carbonate $MgCO_3$, a solid that will precipitate in a solution and can be easily removed. Similarly, calcium hydroxide produces the chemically stable $CaCO_3$, which also precipitates, may be easily removed from a water solution, and sequestered in nearby landfills.

The problem with this method of CO_2 removal is that very high quantities of CaO or MgO are required for the chemical reactions to take place. For example, a typical 400 MW coal power plant emits approximately 11,100 tons of CO_2 every day. In order to convert and then sequester this amount of CO_2, the power plant needs 10,100 tons of magnesia, MgO, or 14,700 tons of calcium oxide, CaO, a very large quantity of mineral to mine, produce, and transport daily to the power plant [14].

CO_2 conversion to solid salts for sequestration requires huge quantities of other minerals that must be mined, processed, and transported to the CO_2 production units. Not only are these schemes costly and consume a lot of energy, but they also cause significant environmental damage with their mining and transportation.

3.10 CONDITIONS FOR ZERO-CARBON EMISSIONS IN THREE SCALES

Some of the approaches and schemes listed in Sections 3.1 to 3.9 may become helpful in the future but cannot be relied on to meaningfully reduce the CO_2 emissions in the atmosphere, either because they consume tremendous amounts of energy and become extremely costly, or because their technology is unproven, or because the risks associated with the technology are substantial. A careful technological and economic examination of all the proposed approaches reveals that the currently feasible solutions to avoid the atmospheric CO_2 buildup and the implied GCC are:

1. Energy efficiency and conservation measures to reduce global energy demand.
2. Significantly higher use of RESs.
3. Increased use of nuclear power including the development of small modular reactors.
4. Conversion of coal power plants into natural gas. This is considered a short-term step, until a zero-fossil-fuel solution is achieved.

All four approaches entail the development of significant infrastructure and require high investment.

From the beginning, it must be emphasized that the global journey to a zero-carbon future will not materialize by adopting a single one of the mitigation approaches. Depending on the available primary energy sources, the state of technology, and available capital for investment, regions and nations will most likely adopt a basket of approaches and will try to apply them in an optimal manner. The Scandinavian countries cannot rely so much on solar energy, but they have good wind and hydroelectric resources (Norway) and the technology to build nuclear SMRs. The islands in the Pacific Ocean can adopt solutions that include both wind and solar energy, but (at least in the near future) cannot realistically rely on nuclear reactors. Conservation, efficiency, and optimization of energy systems are options for everyone to adopt. Actually, most engineering studies and roadmaps agree that it is highly advantageous for conservation and energy efficiency measures to be developed before any investment for infrastructure in renewables or nuclear energy [10, 12, 48]. In the following subsections, we will examine the systems and infrastructure needed for the decarbonization of a single building, of an entire community of 50,000 homes, and of an entire electricity grid.

3.10.1 Net-Zero and Grid-Independent Buildings

Solar PV panels and their circuits are modular, come in many sizes and power ratings that can be combined, may be easily installed, and are passively operated. On the contrary, all wind turbines are bulky, and they are more efficient when they are large with power ratings higher than 1 MW – a power level too high for the typical home or small building. For this reason, PV panels are the only option for single buildings to generate electric power from renewables.

It is common practice to install solar panels on the roofs of homes and small buildings, so that part or all of the electricity consumed in the building is generated by solar energy. When numerous panels are placed on the roof (or close to the building) it is possible to generate over an entire year as much electric energy as the building consumes during the year. Such buildings are called net-zero energy buildings (NZEBs) and are touted as ideal to reduce carbon emissions and mitigate the GCC at the small scale of single buildings. All NZEBs are connected to the electricity grids, they supply the grid with electricity during sunny days and rely on other power plants to supply the electricity they need when the sun does not. For the average home in Texas, which consumes 13,128 kWh per year, to become a NZEB, it would need the installation of approximately 6.2 kW of PV cells. A typical building that has already been refurbished to be energy efficient would require only 4.5 kW of installed PV power [49]. Since the NZEBs draw all the energy needed from the grid during the nighttime and on cloudy days, they do not need energy storage systems.

While NZEBs contribute to decarbonization, they do not have the capability to store the excess energy they generate, and the excess energy is supplied to the grid. During the evenings and nights, they draw all their energy demand from the grid. The proliferation of NZEBs is equivalent to constructing large numbers of PV solar farms and contributes to the development of the duck curve in the demand for the grid (Section 3.3.2 and Figure 3.3). In the long run, NZEBs will have to include energy storage to store the excess energy they generate during sunny days and use the stored energy during hours of insufficient irradiance.

The type of buildings that should be developed in a sustainable society are the grid-independent buildings (GIBs) that are constructed with sufficient energy storage capacity to supply their energy needs at all times without connections to the electricity grid. GIBs are ideal for a sustainable future but require additional investment in their energy storage systems. GIBs also require to generate from the PVs more energy than the total annual building demand, because the storage-regeneration process has limited round-trip efficiency and consumes additional energy. For the average and efficient home in Texas to become a sustainable GIB, using batteries for storage, the installation of 5.4 kW of PV power is required, as well as the installation of a battery-based system with total capacity of 1.8 MWh. If a hydrogen-based storage system, which has lower storage-regeneration efficiency, is to be installed both the required power and storage capacity increase [51].

3.10.2 A Net-Zero Carbon Community

A substantially large zero-carbon community of buildings/homes has the distinct advantage that it may seek an optimal combination of RESs to satisfy its energy needs. Not only PV cells may be used for electricity generation, but also large and more efficient wind turbines at an optimal combination with the PV units. The latter may be placed at a distance from the homes to avoid the turbine-generated noise and the aesthetic offence to some (yes, some people think that wind turbines are ugly and destroy the view from their homes!). If the community is large enough, a small nuclear reactor may be used for the generation of base-load electric power in addition to the renewable units (also constructed at a distance, to minimize any perceived risks).

Synergies emerge in the energy demand patterns of a community that help with the infrastructure. The diverse occupants of the several buildings in the community have different energy usage patterns at the different hours of the day. Some people wake up at 5:00 and start their coffee makers, others at 6:00 or 7:00 am; some cook dinner at 5:30 pm, others at later time. The diverse patterns tend to smoothen the total community electricity demand, and this translates to less energy storage requirements, even if the average energy demand of the buildings remains the same. In addition, buildings in a community may share not only their electricity generation systems, but also their energy storage and regeneration equipment. Essentially, such futuristic communities may share their entire electricity infrastructure and form self-sufficient microgrids. Several such microgrids may be connected to help each other in case of emergency by providing some energy or hydrogen (if hydrogen is chosen as the energy storage medium).

A recent study examined the infrastructure required for such a small, self-sufficient microgrid of 50,000 residential homes in the North Texas region [31]. Solar PV panels and wind turbines provide the needed electric power with the option to include nuclear energy in the form of an SMR. Figure 3.9 depicts a schematic diagram of the essential systems and their interconnections for such a small microgrid, which is powered by a combination of solar PV panels and a number of wind turbines. The inverters convert the DC current to AC current to be used in the homes. The energy storage system may be a large array of battery systems, hydrogen tanks, or an optimal combination of the two. The hydrogen storage systems would include

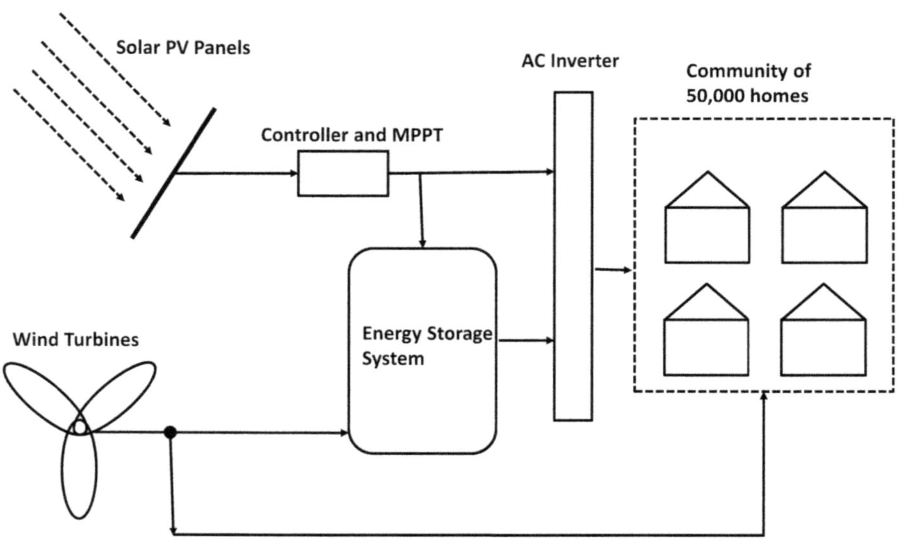

FIGURE 3.9 Elements of a microgrid powered by solar and wind power that supplies with electricity a community of 50,000 homes.

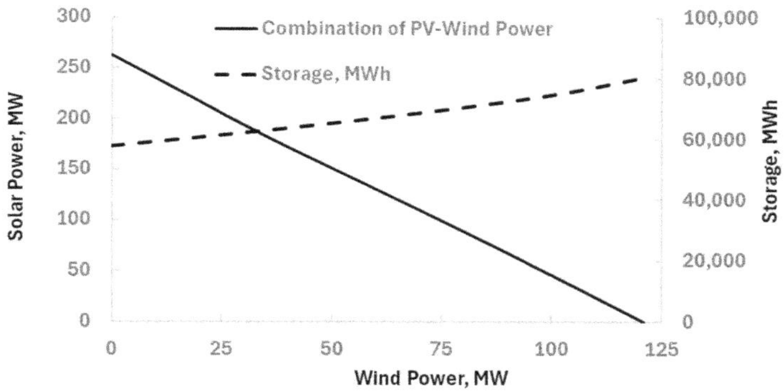

FIGURE 3.10 Combinations of wind and solar power, and energy storage requirements for the microgrid of the community of Figure 3.9.

electrolyzers that generate hydrogen gas from water, compressors to pressurize the generated hydrogen (typically to pressures in the range of 300–700 atmospheres), and fuel cells to re-generate electricity from the hydrogen gas.

Figure 3.10 offers quantitative information on the installed wind and solar power required for the operation of such a microgrid that supplies the 50,000 homes with electricity. It also shows the required storage capacity (scale at right). The ordinate represents the wind power, and the left abscissa represents the solar power capacity needed for the operation of this microgrid. The pair combinations of wind and solar power are depicted in the solid line. For example, the microgrid will function well with a blend of 35 MW wind power and 183 MW solar power, or with a blend of 100 MW wind and 46 MW solar power, or any of the combinations shown by the solid line. The necessary energy storage capacity for each combination of wind-solar capacity is depicted by the dotted line, which reads on the right abscissa. The energy storage capacity increases with the wind power, primarily because increased wind power implies less solar power. In regions of high A/C energy demand, such as North Texas, higher solar power occurs on the hotter summer days, directly provides electricity to the A/C systems, and implies lower need for energy storage.

It must be noted that the addition of a small nuclear reactor, which continuously generates power for this community, would significantly decrease the needed wind and solar power capacity as well as the energy storage needed in the community, as it becomes apparent from a glance at Table 3.1.

3.10.3 A Net Zero-Carbon Grid

A large net-zero carbon grid has not become a reality yet, but there are several computational studies that can give us an idea of the infrastructure needed to develop one. One of these studies pertains to the ERCOT grid that serves 92% of Texas [50]. ERCOT is an extensive grid spread over an area of 250 thousand square miles (630 thousand square kilometers). It transmitted to its customers more than 485 TWh in 2025 (more than the electricity consumed in the UK and Ireland combined).

TABLE 3.3
Required Infrastructure Capacity at Different Scales of Decarbonization

Decarbonization Scale	PV Power, kW	Wind Power, kW	Storage, MWh
Single Home	5.4	0	1.8
Community of 50,000 homes	100,000	46,000	74,500
Entire ERCOT grid	63,000,000	132,000,000	41,600,000

In addition to the wind and solar energy sources, nuclear energy is already contributing to this electricity grid with four nuclear reactors totaling almost 5,000 MW nominal power. More nuclear power plants to generate additional electricity for the grid is a feasible option.

The study on ERCOT determined that with the existing nuclear units continuing to generate electricity at their current levels, for ERCOT to become a zero-emissions grid and satisfy the 2025 electricity demand, it will need an additional 132,000 MW wind power, an additional 63,000 MW of PV cells, and approximately 41.6 TWh of energy storage capacity – either batteries, hydrogen or a combination of the two. By adopting extensive energy efficiency measures by the consumers (which would require additional investment at the household level) it would be possible to reduce the required power by 20–25% and the energy storage capacity by 10–15%. Interestingly enough, this study concluded that it is feasible to substitute all the coal power plants within ERCOT by simply adding solar and wind generation units, and without any additional energy storage capacity. In this case, the existing and flexible gas turbines would be sufficient to satisfy the consumers demand at all times of the year.

Table 3.3 gives a quantitative summary of the infrastructure required for the decarbonization at the three levels presented in this section. It is apparent that significant investment in new systems will be needed for the infrastructure that will lead to a meaningful CO_2 emissions reduction of homes, communities, and entire grids. While the data in this table are results of computational studies and should be considered as approximate, they give a fairly accurate idea of the enormous infrastructure that needs to be developed before complete decarbonization is achieved at any scale of the electricity generation industry.

In summary, it is technologically feasible to meaningfully achieve the transition from fossil fuels to renewables (and nuclear) at the home-level, the community-level, and the grid-level. However, this transition entails substantial investment in infrastructure related to non-carbon emitting electricity generation units, energy storage capacity, and transmission grids.

3.11 THE EFFECT OF ELECTRIC VEHICLES

Electric Vehicles (EVs) have become ubiquitous since the 2010s in all countries. Essentially, the vehicles do not burn any hydrocarbon fuel – and, therefore, do not emit CO_2 – but use the electric energy stored in a high-capacity battery for their motive power. The capacity of the batteries for passenger EVs is in the range 40–100 kWh, and the batteries are charged by connecting them to the grid, typically via a

220–240-volt connection. The proliferation of EVs in several markets worldwide is one of the reasons for the growth of electricity consumption.

While the EVs by themselves do not emit any CO_2, it is wrong to think that they are not associated with greenhouse gas emissions. Electricity is not a primary energy source and must be generated using other sources, as shown in Figure 1.7. If a high percentage of a country's electric power is generated by nuclear or renewable primary sources – e.g., France, Belgium, Norway, Canada, Brazil, etc. – then substituting a gasoline-powered car with an EV avoids a large fraction of the CO_2 that would have been emitted from the internal combustion engine. On the contrary, in countries that generate electricity with a high fraction of coal, the CO_2 avoidance is low and in some of them negative, which implies that substituting an internal combustion engine car with an EV, more CO_2 is emitted (from the power plant that supplies the electricity for the EV battery) [51]. Table 3.4 shows the substitution effect of EVs for several countries based on the fuel mix for electricity generation in 2025 [52]. It is observed that substituting a gasoline-powered car with an EV in South Africa or India, would increase the CO_2 emissions. For the entire world, the CO_2 avoidance is a mere 26%. The EU and Canada, where a great deal of electricity is generated by nuclear power and renewables, enjoy a high level of CO_2 emissions avoidance at 74% and 80% respectively.

The large energy capacity of EV batteries (some approach 100 kWh) may be employed for energy storage in households or (in the case of a large fleet of EVs) for grids. These concepts of energy storage are often referred to as V2H and V2G storage, respectively. The EV batteries may be especially helpful in the transition to renewable energy in regions where people typically use public transportation and their private vehicles remain idle during the workweek (as in most European and Asian large cities). While energy storage in EV batteries will marginally increase the overall efficiency of grid energy storage, it will have little impact on the total storage needed in energy-intensive households and grids, such as those in the USA

TABLE 3.4
Percent of the Annual CO_2 Emissions Avoided by the Substitution of a Gasoline-Powered Car with an EV in Several Countries

Country	% CO_2 Avoidance
Australia	11
Brazil	89
Canada	80
European Union	74
India	−17
Japan	33
P.R. China	4
South Africa	−30
USA	40
Global Average	26

[53]. In addition, the more frequent charge-discharge cycles of the EV batteries will increase their degradation and will void the vehicle manufacturers' warranties. It is characteristic of the EV manufacturers' reluctance for their EV batteries to be used in V2G or V2H applications that in 2025 only two of the more than forty EV models marketed in the USA had V2G capability, with a third having V2H capability.

3.12 PATHWAYS AND ROADMAPS FOR GLOBAL CLIMATE CHANGE MITIGATION

International organizations directly or indirectly related to the United Nations have offered pathways and roadmaps to achieve zero CO_2 emissions by a certain time in the future –typically by 2050 or 2070. The IPCC offered such pathways and recommendations in their reports [10, 11, 49], and the IEA published in 2021 a full roadmap for the transition [12] and two years later suggested an update to this roadmap [54]. A key element in all roadmaps is that CO_2 emissions fall steeply from the almost 37 billion tons in 2022 (39.1 billion tons in 2025) to 24 billion in 2030 and to 13.5 billion by 2035 (a 67% decrease!). Figure 3.11 shows how this reduction in key industries may lead to zero emissions by 2050, according to the IEA.

One notes in this roadmap that the transition of the electricity generation sector to non-carbon primary energy sources (mainly renewables and nuclear) is to be achieved faster than in the other sectors of the economy. At the end of the transition period (year 2050) the authors of the roadmap anticipated that the "other" category will capture and sequester 2 billion tons of CO_2 to allow industry and transportation an equal quantity of emissions. Milestones for this transition include the rapid growth of nuclear and renewable energy generating capacity and extend to investments in electricity grids and drastic coal abatement, as shown in Table 3.5.

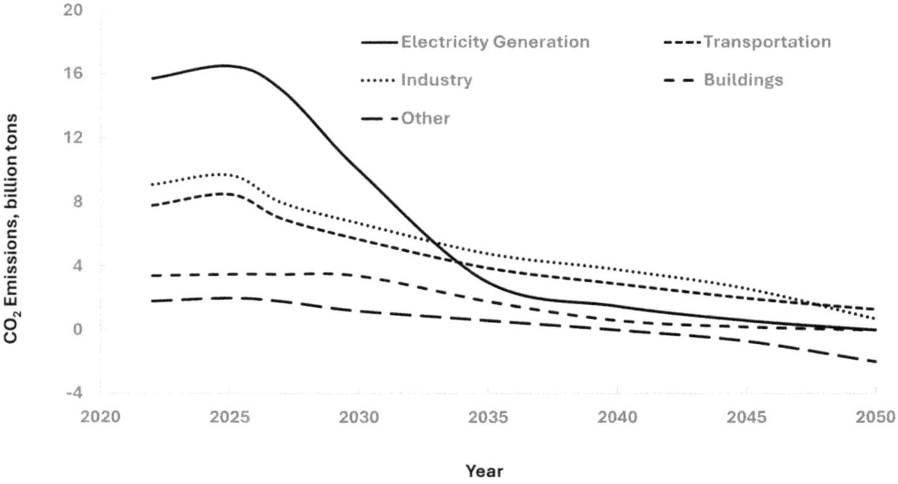

FIGURE 3.11 IEA roadmap (2023 update) to zero-carbon emissions by 2025 for key industries. The actual 2025 data were included for context.

TABLE 3.5
Key Milestones for the Global Electricity Generation Sector in the IEA Net-Zero Energy Scenario

	Renewables, TW	Grids Investment, Billions $US-2022	Nuclear Capacity, GW	Coal, TWh	Natural Gas, TWh
2022	3.5	370	430	10.5	7
2030	10.2	740	560	5.1	6.2
2040	23	1050	850	0	1.5
2050	30	825	910	0	0.3

Source: Data Adapted from Figure 2.15 of Reference [12].

It is observed in Table 3.5 that key parts of the transition roadmap are the tripling of the renewable energy electricity generating capacity, the doubling of the global investment in electricity grids between 2022 and 2030 to accommodate this rapid-pace expansion of renewables, the doubling of the nuclear generation capacity between 2022 and 2040 and the halving of the quantity of coal used for electricity generation between 2022 and 2030. However, it is noted that, at the time of writing this book (2026) no meaningful steps have been taken by the global community, and no national policies were enacted toward achieving these milestones.

3.13 ROADMAPS, NATIONAL POLICIES, AND STATUTES

A closer examination of the milestones to achieve net-zero GHG emissions by 2050 and the connection of such goals to realistic engineering projects for the CO_2 emissions reduction, proves that the current roadmaps are extremely optimistic and unlikely to be followed by the majority of nations. Many of the roadmaps are simple wish-lists. They are based on questionable assumptions and unreasonable expectations to "prove" how complete decarbonization may be achieved in a predetermined timescale. However, they tend to ignore a plethora of technological challenges, such as how the decarbonization of the maritime and aviation industries can become a reality; how to produce more cement, concrete, and steel to house and serve an increasing and more affluent global population; and, most important, how to produce affordable energy for the less affluent segments of the population and for the developing nations.

It is rather unfortunate that the lofty expectations of the roadmaps have been adopted by the UN agencies disregarding the following attributes that characterize sound engineering practice:

a. Availability of capital.
b. Availability of technology and materials.
c. Availability of a reliable supply chain for materials.
d. Development of necessary infrastructure.
e. Prioritization of projects and realistic project timescales.

Tripling the renewable energy generation capacity within eight years (2022 to 2030) requires enormously high investments, especially in the developing world. While several of the OECD countries and P.R. China may raise the capital to construct additional solar and wind farms, most of the other countries, representing more than 70% of the global population, cannot or will not raise the required capital, as it is apparent in the recent investment data of IEA [55]. This reality is reflected in the nationally determined contributions (NDCs) in support of the Paris agreement, where most nations (including the populous India, Nigeria, Indonesia, and Mexico) conditioned their contributions on the availability of financial resources, technology transfer, and the development of capacity-building support. The $300 billion annual investment (to be reached by 2035), promised during COP-29, and very likely with most funds contributed as loans, are not sufficient to fund the necessary projects worldwide, certainly not before the 2030 timeframe. And, if these funds are spent on "loss and damage," their effect on meeting the IEA milestones will be minimal.

Developing sufficient energy storage capacity is a necessary condition for the transition to renewable energy. The timescale for the development of energy storage capacity to support the renewable energy expansion is long and does not fit in the timeline of the roadmaps. The world currently (2026) has less than 5% of the necessary storage capacity, and more than 94% of this existing storage capacity is in PHS systems [24]. The design and construction of these gigantic systems for energy storage takes at least ten years and their local environmental impact is significant. The *Alto Tâmega* PHS project of Figure 3.5 (a medium-range PHS capacity project) took 12 years from design to completion and 550,000 tons of concrete. If the world were to build sufficient PHS capacity for the energy transition its completion will extend well beyond the year 2050. For substantial battery grid-scale energy storage, the quantities of lithium, cobalt, vanadium, and other metals that are essential for the manufacturing of sufficient quantities of battery cells, by far exceed the global mining capacity of the metals, and very likely exceed the global recoverable resources of these metals. Energy storage, which is absolutely necessary for the reliability of the renewable electricity grids of the future, is a traffic bottleneck for this transition, but this is not reflected in the roadmaps and milestones.

Doubling the investment in grids is highly doubtful in the current era of government budget deficits. Reducing by 50% the quantity of coal burned for electricity generation flies in the face of current reality: India and China – two fast-developing economies that account for more than one third of the global population – also account for 62% of the global coal consumption, and coal is currently their main source for electricity generation. It is unrealistic that the two countries will be able to maintain their path to economic development, while phasing out all their coal-generating power plants by 2040. Actually, in September 2025, P.R. China announced that by 2035 it would cut its greenhouse gas emissions by 7–10% from peak levels, without specifying what the peak will be – an action that alarmed environmentalists worldwide. Other highly populated countries, such as South Africa, Turkey, Indonesia, Poland, and South Korea, share similar dependences on their coal power plants and show very few signs to curb their appetite for coal-generated electricity. It appears that, despite the dire projections about the global climate, most of the nations – perhaps with the exception of EU, Japan, and a few others – are unwilling to impose limits on their economic development that would keep global warming below 1.5°C, (or even below 2°C).

Geopolitics and international affairs also play an important role in the timescale for decarbonization [56]. The war in Ukraine caused a short-term spike in the use of coal in Europe and also provided the excuse for several governments to delay their decarbonization investments. The election of the Trump administration in the USA triggered the exit of USA from the Paris accord and ushered the "drill baby drill" era for the extraction of more petroleum and natural gas. The Iran war in early 2026 and the blockade of the Strait of Hormuz significantly affected the crude oil prices, with the result that gasoline prices increased and consumers drove less. After analyzing the higher electricity prices due to more wind and solar units, several political parties in the UK, Canada, and France have adopted a more friendly attitude toward fossil fuels, which may translate in future national policies.

The data from the year 2025 (one third of the pathway between 2022 and 2030) indicate that the celebrated milestones of the IEA roadmap are unlikely to be met: the global capacity of renewables to generate electricity is projected to increase by 50% (and not triple) between 2022 and 2030 [57]. The global consumption of coal increased by 8% between 2022 and 2025 instead of falling. Based on the planned nuclear projects and the global permits for the construction of nuclear power plants, it is highly unlikely that the 850 GW of installed nuclear power will be achieved by 2040.

Combating the GCC and reducing CO_2 emissions is not a pressing priority for most world governments, which tend to legislate with a provincial and not a global mindset. This is indicated by the fact that only 11 of the 195 signatories to the Paris Agreement submitted their NDCs on time in 2025 [58]. The IEA does not have the power to intervene and impose its own rules and have conceded that [56] "...national commitments to reduce emissions collectively fall short of what is required by 2030 to bring global emissions down to a level in line with achieving net zero emissions by 2050. In addition, the various commitments are not yet underpinned by sufficiently strong and comprehensive policies to give confidence that they will be successfully delivered."

National governments have different and pressing economic priorities, other than following an externally imposed zero-carbon roadmap. And most governments do not have the political will to take painful economic measures at present to decrease CO_2 emissions and combat a future GCC, so they "kick the can down the road" and expect those who follow to make the hard decisions and investments for less CO_2 emissions. As a result, it is highly unlikely that the lofty goals of the pathways and roadmaps to achieve net-zero emissions by 2050 will be reached. Actually, the 2025 report by the IEA [59] concedes that "The updated [net-zero emissions scenario] reflects the fact that exceeding 1.5°C is now inevitable..." The opinion of the author of this book is that it is possible that, with thoughtful engineering planning, appropriate financing, and reasonable construction timescales, the global CO_2 emissions will plateau at some point in the future, leading to reductions, and, perhaps, close to net-zero sometime in the late 21st century.

NOTES

1 Bollen J., Manders T., Tong P., 2003, The Kyoto Protocol and EU competitiveness, International Conference on Economic Modeling and Data Science (EcoMod2003), July, Istanbul, Turkey.
2 Lomborg, B., 2001, *The Skeptical Environmentalist*, Cambridge Univ. Press, Cambridge (25th printing, 2016) p. 287.
3 Luhnow D. and Fairness T., 2025, Europe is losing, The Wall Street Journal, August 23.
4 De Leo, G., Rizzi, L., Caizzi, A. et al., 2001, The Economic Benefits of the Kyoto Protocol. *Nature* 413, 478–479 https://doi.org/10.1038/35097156.
5 Data Sources: https://data.worldbank.org/indicator/NY.GDP.MKTP.KD.ZG and https://ourworldindata.org/energy-production-consumption
6 SEER is seasonal energy efficiency ratio.
7 For several practical examples on efficiency and conservation and calculations of the energy savings see Michaelides, E.E., 2018, *Energy, the Environment, and Sustainability,* CRC Press, Boca Raton FL.
8 IEA, 2013, *Energy Efficiency Market Report – 2013,* IEA publications, France.
9 IEA, 2024, *World Energy Investment 2024,* 2024 IEA publications, France.
10 IPCC, 2022, *Climate Change 2022: Mitigation of Climate Change*, (report available from the IPCC website www.ipcc.ch)
11 IPCC, 2014: Climate Change 2014: Synthesis Report. Contribution of Working Groups I, II, and III to the Fifth Assessment Report of the Intergovernmental Panel on Climate Change [Core Writing Team, R.K. Pachauri and L.A. Meyer (eds.)], IPCC, Geneva, Switzerland.
12 IEA, 2021, *Net Zero Roadmap: Net Zero by 2050 A Roadmap for the Global Energy Sector*, IEA publications, France.
13 In 2025 tidal energy generated less than 10 TWh of the 32,000 TWh generated globally. Wave energy and OTEC, while often mentioned in the popular press supplied less than 1 TWh.
14 Michaelides, E. E., 2018, *Energy, the Environment and Sustainability,* CRC Press, Boca Raton.
15 Homer (c. 800 BCE) *Iliad,* Oxford University Press, Oxford, 2011.
16 Freeman E., Occello D, Barnes F, 2016, Energy Storage for Electrical Systems in the USA, *AIMS Energy,* 4, 856–875.
17 California ISO, 2016, Fast Facts – What the duck curve tells us about managing a green grid, report Comm PR/2016.
18 https://www.ercot.com/files/docs/2025/01/18/ERCOT-Monthly-Operational-Overview-December-2025.pdf
19 Michaelides, E.E., 2021, Energy Storage and Installed Wind Capacity Requirements for the Substitution of Fossil Fuels in the Electricity Generation Sector, *Journal of Energy and Power Technology*, **3**(3):18, doi:10.21926/jept.2103040.
20 Marcu A. and Cabras S., 2022, The energy prices crisis and the green transition, quo vadis Europa? ERCST publications, https://ercst.org/wp-content/uploads/2021/11/20211125-Electricity-prices-note-final.pdf.
21 Our appliances and engines need power (in Watts, kilo-Watts or mega-Watts) to operate. Energy (in kWh or MWh) is the multiplication product of power and the time of operation.
22 Reuters 4/30/2025.
23 United States Department of Energy DOE Global ES Database, https://www.sandia.gov/ess-ssl/global-energy-storage-database-home/
24 Michaelides, E.E., 2025, *Energy Storage – Driving the Renewable Energy Transition,* CRC Press, Boca Raton, Florida.
25 Smil, V., 2020, *Energiewende, 20 Years Later,* Spectrum, IEEE publ. December.
26 Rehm TE, 2023, Advanced Nuclear Energy: The Safest and Most Renewable Clean Energy, *Current Opinion in Chemical Engineering,* 39,100878, ISSN 2211–3398, https://doi.org/10.1016/j.coche.2022.100878

27 Data from https://www.nei.org/resources/statistics/us-nuclear-generating-statistics
28 Michaelides E.E. and Michaelides D.N., 2020, Impact of Nuclear Energy on Fossil Fuel Substitution, *Nuclear Engineering and Design*, 366, 110742.
29 Kannan, R., Turton, H., 2016. Long Term Climate Change Mitigation Goals Under The Nuclear Phase Out Policy: The Swiss Energy System Transition. *Energy Economics* 55, 211–222. https://doi.org/10.1016/j.eneco.2016.02.003.
30 Koivunen T, Hirvijoki E, 2024, Effect of Investment Cost on Technology Preference in a Flexible, Low-Carbon Finnish Power System, *Nuclear Engineering and Design*, 417, 112854, https://doi.org/10.1016/j.nucengdes.2023.112854.
31 Michaelides E.E., and Michaelides D.N., 2025, Small modular reactors and the transition to renewable energy, *Nuclear Engineering and Design,* 441, 114207, https://doi.org/10.1016/j.nucengdes.2025.114207.
32 US-DOE, 2025, *Transformative Power Systems,* Office of Fossil Fuels and Carbon Management, *https://www.energy.gov/fecm*
33 Banks M., 2024, ITER fusion reactor hit by massive decade-long delay and €5bn price hike, *Physics World, Nuclear Fusion News,* July 03.
34 Michaelides, E.E., 2021, Thermodynamic Analysis and Power Requirements of CO_2 Capture, Transportation, and Storage in the Ocean, *Energy,* 230, 120804.
35 Michaelides, E.E., 2025, Power and Energy Requirements for Carbon Capture and Sequestration, *Thermo*, **5**, no. 1: https://doi.org/10.3390/thermo5010008.
36 Kharaka YK, Cole DR, Hovorka SD, Gunter WD, Knauss KG, Freifeld BM, 2006, Gaswater-Rock Interactions in Frio Formation Following CO_2 Injection: Implications for the Storage of Greenhouse Gases in Sedimentary Basins, *Geology*, 34(7), 577e80.
37 Song Y, Chen B, Nishio B, Akai B, 2005, The Study on Density Change of Carbon Dioxide Seawater Solution at high Pressure and Low Temperature, *Energy,* 30, 2298–2307.
38 Winson, T.R.S., 1992, The deep ocean disposal of carbon dioxide, *Energy Conv. and Management,* 33, 627–633.
39 Rochon E., 2008, False Hope - Why carbon capture and storage won't save the climate, *Greenpeace International,* Amsterdam.
40 Piementel D., 2003, Ethanol Fuels: Energy Balance, Economics, and Environmental Impacts are Negative, *Natural Resources Res.* 12,127–134.
41 Wang M., Saricks C., and Wu M, 1997, Fuel-Cycle Fossil Energy Use and Greenhouse Gas Emissions of Fuel Ethanol Produced from U.S. Midwest Corn, *Argonne National Lab. Report,* December.
42 Department of Energy, EIA, 2024, https://www.eia.gov/todayinenergy/detail.php?id=63644#.
43 Hurtes, S., Cai W., 2022, *Sacrificing centuries-old trees in the name of renewable energy,* New York Times, September 10. 2022.
44 Putting CO_2 to Use: Creating Value from Emissions, EIA Report, France, September 2019.
45 Paustian K, Lehmann J, Ogle S, Reay D, Robertson GP, Smith P. (2016). Climate-Smart Soils. *Nature* 532, 49–57.
46 See https://www.wri.org/technical-perspectives/insider-further-explanation-potential-contribution-soil-carbon-sequestration-working
47 IPCC, 2019, *Global warming of 1.5°C,* Intergovernmental Panel on Climate Change.
48 Michaelides E.E., 2025, Energy Efficiency and the Transition to Renewables –Building the Communities of the Future, *Energies,* 18(**7**), 1778; https://doi.org/10.3390/en18071778
49 Michaelides, E.E., 2025, Energy Efficiency and Grid-Independent Buildings, *Journal of Solar Energy Engineering.,* 147, 051001-1, doi: 10.1115/1.4068357.
50 Leonard, M.D., Michaelides, E.E., and Michaelides, D.N., 2019, Energy Storage Needs for the Substitution of Fossil Fuel Power Plants with Renewables, *Renewable Energy,* 145, 951–962.
51 Michaelides, E. E., 2020, Thermodynamics and Energy Consumption of Electric Vehicles, *Energy Conversion and Management,* 203, 112246.
52 Data from IEA. https://www.iea.org/reports/monthly-electricity-statistics-overview

53 Michaelides EE, Nguyen VND, Michaelides DN, 2023, The Effect of Electric Vehicle Energy Storage on the Transition to Renewable Energy, *Green Energy and Intelligent Transportation*, 2(1), 100042.
54 IEA 2023, *Net Zero Roadmap: A Global Pathway to Keep the 1.5°C Goal in Reach 2023 update*. IEA publications.
55 IEA (2024), *World Energy Investment 2024*, IEA, Paris https://www.iea.org/reports/world-energy-investment-2024.
56 Meckling, J, 2025. The Geoeconomic Turn in Decarbonization, *Nature* 645, 869–876. https://doi.org/10.1038/s41586-025-09416
57 IEA 2025, https://www.iea.org/energy-system/renewables
58 Weak plans mean a bleak future': Most countries miss deadline to submit new national climate plans, *Euro News,* February 2025.
59 IEA, 2025, *World Energy Outlook 2025,* IEA Publications, November 2025, report available at www.iea.org.

4 Sustainability Goals and Cost of Decarbonization

It is apparent that all the roads toward significantly curbing the global CO_2 emissions and mitigating the Global Climate Change (GCC) pass through the transition of the electricity generation industry from fossil fuels to renewables. The several plans and roadmaps promulgated by the IEA and the IPCC make this route explicit and definite. For this reason, it is important to understand the monetary cost of the proposed transition and the implications of this cost for the several economic layers of our society. It must be emphasized from the beginning, however, that calculating future prices and costs is not as accurate and precise as calculating the technical parameters for the transition, such as solar and wind power generating capacity, energy storage capacity, dissipation, etc. Fuel prices, commodity prices, and the costs of power generation and energy storage systems are highly variable. As a result, all calculations of the future cost of electricity are laden with high volatility and uncertainty.

The several types of energy subsidies and of the carbon taxes are presented in this chapter, and their impact is examined on the energy markets, the global electricity prices, and the poorer segments of the population [1]. The economics of the decarbonization and the United Nations Sustainable Development Goals are also critically examined, with particular emphasis on the Goals that are pertinent to energy use, climate change, renewable energy, poverty, and inequality. The several elements of the necessary infrastructure for a global decarbonization of the electricity industry are presented (including new solar and wind energy units, land procurement, energy storage systems, and electricity grid overhaul). The cost for building this infrastructure for decarbonization is calculated with reference the electricity grid that serves Texas. It is concluded that the decarbonization investment will have a substantial effect on the electricity prices of the future, which will have to increase by a factor between 3.3 and 4.8, significantly affecting the poorer part of a nation's population and entire nations of the developing world.

4.1 THERMODYNAMICS AND ECONOMICS – VARIABILITY AND UNCERTAINTY OF PRICES

The science and technology of electricity generation is based on thermodynamics, which is a branch of physics. Thermodynamics is founded on physical principles – the laws of thermodynamics – that have been proven to be precise, accurate, and invariant with time. All the physical and experimental observations of the last three centuries have verified the accuracy of the laws of thermodynamics, on which the concept of energy is based. After more than 250 years of experimentation, the numerical values of thermodynamic properties have been accurately determined – with very low uncertainty. For example, the heat to vaporize 1 kg of water and generate steam at

DOI: 10.1201/9781003674658-4

atmospheric pressure is 2,258 kJ per kg of water. The uncertainty of this value is less than ±0.015%, and it is universally accepted that this numerical value will not change outside of its narrow uncertainty limits at any time in the future as it has not changed in the past. Everyone may count on the fact that the latent heat of water (and all material properties) will be the same after thirty years and after three centuries. However, the same cannot be said about the cost of the energy of 2,258 kJ to vaporize this 1 kg of water, which has had in the past and will have in the future a high degree of volatility.

Economics is a social science, based on empirical principles, which are related to geopolitical events, everyday opinions, and behavioral observations of individuals, groups, and nations. The principles of economics emanate from correlations of data. These data are not permanent, actually they are ephemeral; and they heavily depend on the time of observations and the type of national economies– e.g., the energy price mechanisms are different in free-market economies, regulated economies, and centrally planned economies. The demand-supply equilibrium, which is central in economic theory and is said to determine prices and costs, is highly influenced by a variety of factors, including political events, technology, the shifting preferences of consumers, and even fashion. The prices of all forms of energy reflect the abundance or scarcity of the energy sources relative to the demand, the local and global markets, as well as financing.

Unlike the properties of thermodynamics, which are time-invariant, the prices of fuels, materials, equipment, and machinery are not determined by a physical law but follow geopolitical events and supply-demand considerations. The numerical values of economic variables, such as prices, interest rates, etc., are time-dependent; they fluctuate daily and exhibit high variability with time. As a result, all economic predictions and projections of prices and costs in the future are laden with significant uncertainty.

Prices associated with energy projects – e.g., fuel prices, materials for energy storage, and interest rates for energy projects – show higher volatility than those of other commodities, because (in addition to all other effects) the economics of energy are subjected to geopolitical and military developments [2]. An example of the significant fluctuation in energy prices is shown in Figure 4.1, which depicts the nominal and real prices of gasoline in the USA (annually averaged) in the period 1980 to 2026 – a period similar to the life-cycle of electricity-generating power plants [3]. The 2026 data are for the month of March 2026 and reflect the effect of the Iran war. The real prices are expressed in US$ of 2025. A few of the geopolitical events that have substantially influenced the gasoline prices are also pointed out in Figure 4.1.

It can be easily seen in Figure 4.1 that both sets of prices exhibit high fluctuations during this period of 46 years. An analysis of the data shows that, while the two averages of the nominal and real prices are \$1.99 and \$3.24 per gallon, respectively, the percentage standard deviation of the nominal price is 54% and that of the real price is 39% – two values that correspond to high fractions of the average prices. It is evident that any predictions for the price of gasoline on a future date, based on the past prices, will be laden with high uncertainty and can only be a rough estimate. It must also be noted that, if the more volatile daily or monthly prices were selected instead of the annually averaged prices, both values of the fractional standard deviation increase substantially, which implies higher uncertainty for calculations of future prices.

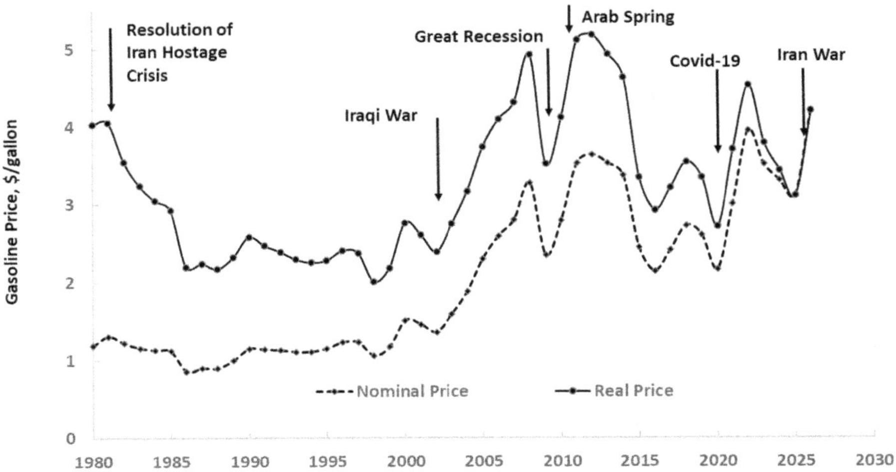

FIGURE 4.1 Fluctuations of the annually averaged nominal and real (in 2025 US$) prices of gasoline in the USA between 1980 and 2026. The 2026 data are for the month of March 2026.

The high standard deviation and high uncertainty of prices are not confined to gasoline but are characteristic of all the fossil fuel prices – natural gas, anthracite, lignite, etc. High price variability also characterizes the materials (commodities) used for the manufacturing of energy storage systems. Figure 4.2 depicts the price of lithium carbonate (the mineral from which lithium metal is extracted) in the period 2017–2025 in Yuan per ton, the currency of the P.R. of China [4]. In the

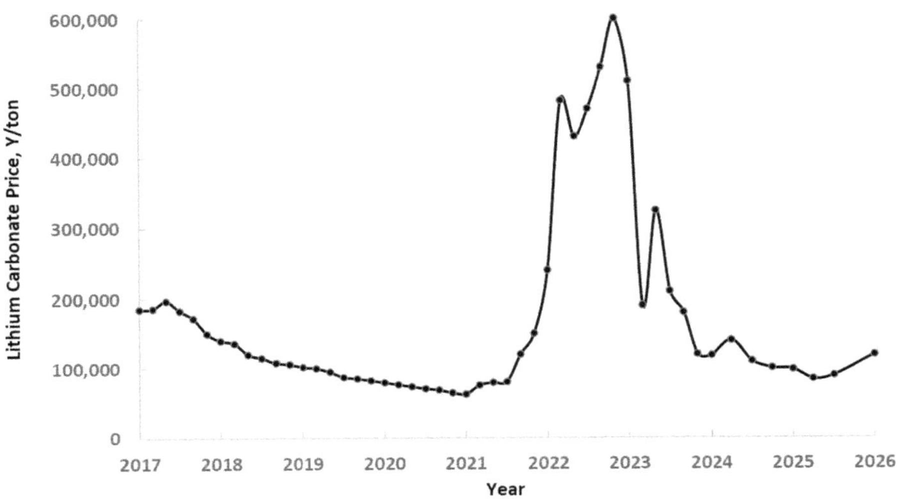

FIGURE 4.2 Monthly-averaged price fluctuations of lithium carbonate in the period 2017–2025.

period 2017–2025, the price per ton of this mineral fluctuated between 50,000 Y and 591,500 Y, with the standard deviation being 84% of the average price. It is apparent from this Figure that any reasonable prediction of the future trends for the cost of lithium carbonate and lithium batteries is laden with high uncertainty.

Because of such high uncertainties in calculating future prices, the best we can do is to make an estimate of the most likely range of future costs associated with all energy commodities and systems. The more distant the future is, the wider this range will be.

4.2 UNITED NATIONS AND SUSTAINABLE DEVELOPMENT

The term *sustainable development* was coined by the UN *World Commission on Environment and Development* (WCED), which was established in 1963 and discharged in 1987. The Commission – also known as the *Brundtland Commission* in honor of its chair, the former Norwegian Prime Minister Gro Harlem Brundtland – established a set of guidelines for nations to achieve higher standards of living (economic development), while taking under consideration the interrelationships between people, resources, and the natural environment [5]. The WCED recognized that the human environment does not exist as an isolated entity, separate from human actions, ambitions, and needs; that economic development is what humans pursue, attempting to improve their conditions within their living habitat; and that the environmental, social, and economic factors, which affect all nations and human communities, are highly interconnected and inseparable. Consequently, the Commission's work shifted the focus of the international community toward integrating environmental and economic policies and laid the groundwork for subsequent international agreements on the environment.

The Commission defined sustainable development (*sustainability*) as the ability of a society to *ensure that it meets the needs of the present, without compromising the ability of future generations to meet their own needs.* In this context, sustainability implies limitations imposed by the current state of technology, recoverable resources, social organization, and the ability of the natural environment to absorb the impacts of human activities. After the publication of the Commission's report, *Our Common Future* [5], the sustainability concept has been adapted and used within several disciplines – e.g., biology, architecture, economics, environmental science, engineering, sociology – in a variety of contexts and connotations, often encompassing the notions of: political empowerment, economic equity, effective citizen participation in the national decision-making processes, greater democracy, and equity in international decision-making fora.

4.3 SUSTAINABLE DEVELOPMENT GOALS, TARGETS, AND INDICATORS

Following the 1987 report on *Our Common Future,* the 1992 conference on Environment and Development (in Rio de Janeiro) introduced the environment and specifically the GCC ("global warming" at the time) as a subject of immediate concern for the global community, which needs to be addressed by concerted and urgent action. The annual Conferences of Parties (COPs), which specifically deal with the GCC, are successors of the 1992 Rio conference and the United Nations Framework Convention for Climate Change (UNFCCC), which took place in New York in the same year. Twenty years later, during the United Nations Conference on Sustainable Development (also known as Rio+20) it was proposed that a set of Sustainable Development Goals (SDGs) be adopted to define sustainable development and provide a blueprint for global action. Consequently, 17 sustainable development goals were proposed and adopted by the UN General Assembly in 2014. Among the key subjects in the adopted goals are the eradication of poverty, modern energy, clean water, education, sanitation, health, and safe human settlements. The seventeen goals in the UN sustainability framework are:

1. End poverty in all its forms everywhere.
2. End hunger, achieve food security, improve nutrition, and promote sustainable agriculture.
3. Ensure healthy lives and promote well-being for all ages.
4. Ensure inclusive and equitable quality education and promote lifelong learning opportunities for all.
5. Achieve gender equality and empower all women and girls.
6. Ensure availability and sustainable management of water and sanitation for all.
7. Ensure access to affordable, reliable, sustainable, and modern energy for all.
8. Promote sustained, inclusive and sustainable economic growth, full and productive employment and decent work for all.
9. Build resilient infrastructure, promote inclusive and sustainable industrialization, and foster innovation.
10. Reduce inequality within and among countries.
11. Make cities and human settlements inclusive, safe, resilient, and sustainable.
12. Ensure sustainable consumption and production patterns.
13. Take urgent action to combat climate change and its impacts by regulating emissions and promoting developments in renewable energy.
14. Conserve and sustainably use the oceans, seas, and marine resources.
15. Protect, restore, and promote sustainable use of terrestrial ecosystems.
16. Promote peaceful and inclusive societies for sustainable development, provide access to justice for all.
17. Strengthen the means of implementation and revitalize the global partnership for sustainable development.

Effectively, the original and simple sustainability dictum for the global society to *ensure that it meets the needs of the present, without compromising the ability of future generations to meet their own needs,* has been replaced by a set of seventeen goals that cover poverty, economic equality, climate, energy, health, justice, empowerment, global partnerships, and several other concepts. Altogether they make a set of lofty goals that will undoubtedly benefit humanity. Every one of the seventeen SDGs is complemented with eight to twelve "targets" that need to be met by the global community, for a total of 169 targets [6].

It is apparent that the seventeen goals were developed by diverse groups of individuals (committees) with each member of the group contributing what they understood as essential elements of a future prosperous global society. However, it is rather troubling that a goal to combat global overpopulation is not directly included in the seventeen goals and the 169 targets. One may argue that it is somehow implied in the documents and some of the actions to be taken. However, an increased population on the planet is already straining the natural resources; has accelerated the atmospheric CO_2 accumulation; has enhanced the adverse effects of the GCC; and is contributing to the national and global economic inequities. Overpopulation will augment all the GCC problems of the future and is a serious impediment to any kind of sustainable development. As such, combating overpopulation deserves its own SDG and targets.

4.3.1 Sustainability Indicators

Sustainable development extends well beyond the principles of science and encompasses multi-layered social, environmental, and economic issues. Motivated by the dictum "you cannot manage what you cannot measure," several nations have defined indicators (metrics) that measure their progress toward the SDGs. The UN has adopted a set of 231 global indicators to monitor the annual progress toward the 17 SDGs and their 169 targets. Every indicator, which is represented by a single numerical variable, provides a quantifiable way to measure progress toward the goals. Also, each indicator is assigned to a *custodian agency* of the UN to develop the measurement methodology, collect the data, and report on progress. Regarding SDG 7 (on sustainable, reliable and modern energy) some of the indicators are [7]:

- Proportion of population with access to electricity.
- Proportion of population with primary reliance on clean fuels and technology.
- Energy intensity measured in terms of primary energy and GDP.
- International financial flows to developing countries in support of clean energy research and development.
- Renewable energy production (including hybrid systems).
- Installed renewable energy-generating capacity per capita.
- Investment in energy efficiency.

The member nations report to the UN on their progress toward sustainability, and the UN has issued annually *The Sustainable Development Goals Report,* since 2015, to disseminate progress in achieving each goal [8].

4.4 FOUR GOALS RELATED TO GLOBAL CLIMATE CHANGE AND POVERTY

Without any doubt all seventeen sustainability goals will improve life on planet earth. If the human race manages to achieve the goals, the future generations will have a happier life than we or our antecedents had. To use a colloquial expression in the USA, the sustainable development goals are "like motherhood and apple pie." No one with common sense will argue against them. The big questions are how we will achieve them, how much will they cost, and what is the path the human society may follow to reach the goals' destinations. In the following sections we will examine how four of the SDGs are interrelated and how aiming for one will have a detrimental effect on others. The four goals to be analyzed are:

1. Ensure access to affordable, reliable, sustainable, and modern energy for all (SDG 7).
2. Take urgent action to combat climate change and its impacts by regulating emissions and promoting developments in renewable energy (SDG 13).
3. End poverty in all its forms everywhere (SDG 1).
4. Reduce inequality within and among countries (SDG 10).

4.4.1 Terminology of the Goals

A cursory look at the seventeen SDGs proves that – perhaps because a broad consensus was needed – the sustainability goals are a little vague and have not been articulated in scientific or legal terms. Regarding the goals on energy and the environment, one may be a little pedantic and point out that all the currently used renewable energy forms (SDG 13) are not modern (SDG 7), actually have been well known and used by humans for millennia. Since the onset of human civilization, solar energy has been used to dry foodstuffs, to dry clothes, to heat the water, and to heat human dwellings. Archimedes is even purported to have used solar energy for military purposes – to burn Roman ships during the siege of Syracuse. Wind energy has been used for at least three millennia to propel sailing boats and generate power in windmills. There are numerous watermills in Europe, China, Japan, Iran, and several other countries. The watermills were used since at least the early Middle Ages to develop mechanical power – in the same way hydraulic energy is now used to generate electricity. And when it comes to geothermal energy, hot springs have been used for millennia by the Greeks, Romans, Japanese, and others for recreational baths and home heating. Coal, petroleum, natural gas, and nuclear fuels are significantly more "modern" energy forms than renewables.

Another observation is that the first part of SDG 10 (reduce inequality within and among countries) can be fast realized by governmental *fiat* – by heavily taxing the wealthy and offer substantial subsidies to the poor, until the incomes of all the citizens become roughly equal. In most countries (unless the average income is below the poverty level) this action will also eliminate poverty, so that SDG-1 will be automatically realized. However, our experience has shown that such an approach will also discourage people from working hard, innovating, obtaining an education, taking risks to improve their economic situation, and (why not?) become wealthy. In extreme cases such an approach may discourage individuals from working at all. In the mid-20th century, the former Soviet Union and several of its satellite countries had effective policies and taxation systems that significantly reduced inequalities within their boundaries. However, this economic model has failed, and its supporters have turned to more market-oriented (capitalist) principles. Even in P.R. China (one of the countries with the least inequality), Chairman Deng Xiaoping declared that "to get rich is glorious." Clearly, the world of the 21st century will have to choose a better path than heavy taxation to eliminate poverty and minimize inequality.

4.5 ENERGY SUBSIDIES

The vague articulation of the SDGs notwithstanding, the main direction of the goals associated with energy and the climate is quite clear: the promotion of renewable energy sources, with the aim to substitute all fossil fuels in the future. The several targets of these two goals and the associated metrics (sustainability indicators) make this aim abundantly clear. To comply with the goals and achieve their targets, the national governments that subscribe to the UNFCCC have adopted appropriate measures in support of renewable energy sources. The most effective among the measures are mandates, regulations, subsidies, and taxes.

In market-oriented economies, where governments cannot dictate the direction of the economy and markets, governments invariably influence the direction of economic sectors by subsidies and targeted taxes. While there is a plethora of ways to subsidize an economic sector, the most common types of subsidies in the energy sector may be classified in the following categories [9]:

1. *Direct expenditures* involve the transfer of funds to recipients by national agencies. For the electricity generation sector, examples include fixed payments per MWh of electric energy generated (e.g., $/MWh), payments per MW of power installed and connected to the grid (e.g., $/MW), and payments for the production of the primary energy sources that generate electricity (e.g., $/ton of coal or $/Mccf of gas).
2. *Governmental mandates* require branches of the government (e.g., defense installations, housing, administration buildings, etc.) to acquire and use designated energy resources. In the third decade of the 21st century, such mandates typically promote renewable energy utilization.
3. *Tax expenditures* include tax credits and tax refunds for the investment in electricity generation projects, such as the construction of solar and wind farms. Energy efficiency measures that promise lower energy consumption – e.g., the installation of efficient heat pumps and air-conditioners in households – also enjoy tax expenditure subsidies in several nations.
4. *Research and Development (R&D) expenditures* include funds paid to industry, universities, and individuals to conduct basic and applied research on designated areas. The support of the basic R&D expenditures is not confined to specific energy forms. They broadly support research on all forms of energy and may include fundamental research on topics peripheral to energy production and utilization, such as high-speed computing, quantum physics, and basic chemistry. The applied R&D expenditures include technological advances in specific energy forms as well as technology for higher efficiency and improved materials.
5. *Loan subsidies* come in governmental guarantees for commercial loans that an energy corporation may take. Loan guarantees – recently applied to renewable and nuclear power projects in the USA – effectively help reduce

TABLE 4.1
Total Cash Subsidies Paid by the USA Federal Government in the Period 2016–2022, in Millions US$

	Direct	Tax	R&D	Total	Energy Produced, EJ	Subsidy per EJ, $million/EJ
Coal	–445	13,324	2,383	15,262	99.4	153.54
Gas and Petroleum	915	7,622	739	9,276	456.8	20.31
Nuclear	172	999	1,684	2,855	61.4	46.50
Renewables	3,127	78,332	2,364	83,823	56.7	1,478.36

the interest rate paid by the energy corporations for the development of new projects. In all the cases of loan guarantees the governments expect that the loans will be fully repaid by the corporations. Loan guarantees are only paid by the government and become subsidies, when energy corporations default and are unable to repay the loans.

It must be noted that all subsidies emanate from specific legislation and can be very complex in language, scope, and intended functions. In the energy sector national governments typically use a basket of subsidies to promote the use of specific energy forms. Since the beginning of the 21st century renewable energy and energy efficiency projects have been heavily promoted in nations that are signatories to the UNFCCC and received the lion's share of the monetary subsidies. Table 4.1 shows the total cash subsidies paid by the Federal Government of USA during the period 2016–2022 (funds in millions US$). Governments of several states within the USA have paid additional subsidies [9].

The negative value in the direct expenditures for coal reflects a "de-obligation of funds" in 2016 and 2017 [9]. It must be emphasized that Table 4.1 only includes subsidies that are directly offered to energy businesses in the USA, as they are defined by the US Department of Energy [9].

4.5.1 Notes on Energy Subsidies – Definitions and Externalities

The underlying philosophy behind energy subsidies is that the technologies of certain energy forms are either in their infancy or not completely developed. For renewable energy – especially wind and solar – a large part of the subsidies is awarded because they are considered to be environmentally benign and do not emit any CO_2. Because the technologies of harnessing wind and solar energy are still considered in the developing stage, subsidies assist corporations that develop technologies for these two resources to become competitive in the energy markets and to fully commercialize their technologies. Essentially, national and regional governments step in to assist with new energy technologies and channel funds to the development of primary energy sources that are cleaner and more sustainable than fossil fuels.

It is expected that, as the new technologies of energy industries develop (and the technologies become mature) the subsidized energy forms will become economically viable and that the governmental support will expire. If the governmental support does not have a sunset date in the future, the continuation of subsidies may create a group of privileged businesses that are too heavily dependent on governmental grants to improve their technologies, lower their cost, and remain viable in an otherwise competitive market. In such cases, the privileged subsidized businesses will lobby politicians and governments to maintain their subsidies; and this leads to a perennial process that wastes taxpayer money and distorts the energy markets.

A significant problem in assessing the recipients and effects of subsidies is the definition of what constitutes a subsidy. For example, are energy-related assistance programs for poor households subsidies for fossil fuels? In the USA the *Low-Income Home Energy Assistance Program* helps millions of poor households annually to pay their energy bills, by distributing billions of dollars to low-income households. Such payments should not be considered as subsidies for fossil fuels simply because high percentages of electricity and heating for low-income families are generated from fossil fuels. However, the authors of several reports and studies have included this type of energy assistance to low-income households as fossil fuel subsidies [10].

Authors of some studies go even further and extend the definition of subsidies to environmental, health, and societal costs that may be related or attributed to the use of certain energy forms, typically fossil fuels. The estimated costs of air pollution, temperature rise, road accidents, and health problems associated with GCC are included in these studies as subsidies for fossil fuels. Not surprisingly, the estimates for such fossil fuel subsidies skyrocket. Bloating the subsidy definitions by including almost impossible to quantify externalities only serves to "muddle the waters," and confuse the public and policymakers in order to encourage them to increase the support for additional subsidies to renewable energy corporations and more intervention in the energy markets.

From the scientific point of view there are also methodological problems with subsidies based on economic externalities such as environmental pollution, health costs, and societal costs. There is not a unanimous agreement of what the social costs are, and there is an order of magnitude disparity of what are the values of these costs. Health costs are based on statistical data of the past to estimate projected illnesses and population mortality in the future; they are also subject to extremely high (by orders of magnitude) uncertainty [10]. Environmental costs are based on estimates to remediate the environment to a previous state – typically called "pristine" state – and may run to hundreds of billions of dollars. There is also very high uncertainty in the estimates of environmental costs and not everyone agrees what is the previous pristine state of the environment. For example, if the previous states of cities such as New York, London, and Shanghai are taken in the 1880s, do we account for the tons of horse manure accumulated in the streets of the cities because there were not internal combustion engines? If the previous state of Florence, Milan, and Paris is in the Middle Ages, do we account for the raw sewage that was flowing in the middle of the streets? The previous states of the environment in the cities where most of us live were not all that pristine in past times. Actually, in the 20th and 21st centuries with the help of abundant energy, humanity enjoys a cleaner environment and significantly more hygienic dwellings.

It must be emphasized that Table 4.1 does not include any externalities stemming from subjective interpretations that are laden with extremely high uncertainty. It only includes tangible benefits – governmental payments, tax breaks, R&D expenditures, and the benefits of loan guarantees – as defined by the USA Department of Energy [9]. The calculation of these tangible benefits does not support the frequently advanced claims that fossil fuels and nuclear power receive more federal subsidies than wind and solar energy.

4.6 ENERGY TAXES – THE CARBON TAX

If subsidies are the carrot to induce citizens and electricity-generating corporations to invest in renewables, energy taxes and carbon taxes are the sticks to discourage citizens from using excessive quantities of energy and deter corporations from investing in projects that emit CO_2. Energy taxes, which target electric power regardless of its source, are imposed at the consumer level and aim at encouraging energy conservation and increased efficiency in households. Expensive electricity makes higher efficiency projects – e.g., household insulation, LED lighting, efficient air-conditioners, etc. – more profitable and agreeable to the consumers. When, as a result of higher efficiency, the consumption of electricity decreases in a nation, the CO_2 emissions follow suit. At the same time, since carbon taxes are directly aimed at CO_2 producers, the taxes compel CO_2 emitters to make their current processes more efficient and to eventually transition these processes from burning fossil fuels to using renewable energy sources.

Carbon-use barriers are imposed in two forms: the first is "cap-and-trade" and the second is a direct tax on CO_2 emissions (e.g., $ per ton of the gas emitted). Both of them may apply to the other GHG gases (e.g., methane and refrigerants). The taxes on the different gases may vary to reflect the national priorities and the relative contribution of the individual gas to GCC.

The cap-and-trade system has been successfully implemented in the European Union (EU), since 2005. It imposes an emissions cap (upper ceiling) on several industries, including power generation, airlines, and cement. CO_2-emitting corporations receive emissions allowances (e.g., 300,000 tons of CO_2 emitted for a given year) or purchase from other corporations such allowances. The total quantity of the GHG allowances is capped at a national level that decreases annually. Corporations that implement substantial GHG reduction projects may sell their excess allowances to others and profit from their projects. With the annually decreasing caps, it is reasonably expected that GHG emissions will reach the desired low levels within a predetermined year.

The imposition of a carbon tax is based on the "polluter pays" principle and is a lot simpler. Corporations that emit GHGs pay a tax or penalty, which is proportional to the type of gas and the amount of gas emitted. For example, electric power corporations that burn lignite to generate power pay a given tax per ton of CO_2 emitted. In most cases, this tax increases with time, for the emissions to fall within the guidelines and expectations of the national governments. Among the most stringent national carbon taxes, the carbon tax in Sweden was introduced in 1991 at 250 Kronen ($24) per ton of CO_2 emitted and gradually increased to 1,510 Kronen ($140) by 2025 [11].

Since carbon taxes are eventually passed to the consumers as higher prices, governments may mitigate their effects on the economy and use the carbon tax revenue, as well as any cap-and-trade revenues in a variety of ways, among which the most popular are:

1. Lower payroll and income taxes, to mitigate the negative effects of high energy costs on workers and small businesses.
2. Fund climate initiatives, such as research and development efforts for clean technologies, climate change mitigation projects, and adaptation projects.
3. Use the funds to improve public services and infrastructure, such as education, public transportation, basic research, and services that promote economic growth.
4. Reduce the overall national deficit or the national debt (for debtor nations).
5. Distribute all or part of the revenue to lower-income households via direct grants or dividends, which will help offset the higher energy costs.

In most countries that have adopted a carbon tax, policymakers combine more than one of these methods to optimize the taxation impacts.

The strict imposition of GHG cap and trade policies and the carbon taxes have had some undesirable consequences. Cap and trade policies within a group of nations have resulted in the so-called *carbon leakage* that affects national industries. Carbon taxes – whose cost is eventually passed by the corporations to the consumers – if they are not well managed, they can be *regressive* and increase poverty and inequality within a nation.

4.6.1 Carbon Leakage

Countries across the globe have different GHG emissions reduction targets and policies. Several nations have not adopted carbon taxes, a policy that confers significant advantages to energy-voracious industries within their borders. The term *carbon leakage* is a general term that applies to corporations and industries leaving a country with stringent GHG emission standards for another country, where such standards are lax or non-existent. The irony in this situation is that corporations, which have migrated from country A with high GHG standards to country B with lower or non-existing standards, may manufacture products in their new country B, without having to pay carbon taxes, and export their products back to country A at lower prices. This practice significantly disadvantages the industries that remained in the (high carbon tax) country A and threatens them with extinction in the long run.

The EU, which has adopted stringent decarbonization standards since the beginning of the 21st century (and has achieved a 33% emissions reduction in the period 2005 to 2021) has been particularly affected by the adverse effects of carbon leakage. Corporations in the cement, aluminum, steel, fertilizer, and similar industries have left European jurisdiction territories to establish themselves in countries with lower GHG emissions standards. The same corporations export their products at highly competitive prices back to the European consumers.

To counteract the trade imbalance, the European Parliament adopted a carbon border adjustment mechanism (CBAM) in 2021, with implementation date January 2026 [12]. Essentially, the CBAM puts a price on the GHG emissions associated with imported goods and imposes commensurate border tariffs. The intent of the CBAM is to level the playing field between the European industries and foreign producers in countries that have lower environmental standards. Proponents of the CBAM claim that the border tax will increase the competitiveness of the European industries; will facilitate the adoption of more rigorous environmental standards in other countries; and may even induce the run-away industries to come back to Europe. Opponents of the tax claim that it may hurt the existing European industries by making imported parts more expensive, thus inducing consumers to buy the finished product from another country – e.g., if the imported steel costs more, European cars and wind turbines will become more expensive, and the prospective clients may shift to buying these entire systems from a non-EU country.

Implementing the CBAM by the EU authorities will also meet obstacles, chief of which is the shock to the precarious equilibrium of the global trade environment. Trade tariffs tend to be reciprocal, and the imposition of the CBAM may result in a backlash and new tariffs on European exports. There are already (early 2026) allegations that the imposition of this mechanism violates the General Agreement on Tariffs and Trade (GATT) and that it will significantly harm the global trade [13]. A second obstacle to CBAM is related to the accuracy of the calculation of the tariff, given the global nature of supply chain and global value chain of materials. Materials in a final product often come from a group of countries with entirely different environmental standards. It is very difficult (almost impossible) to determine what the contribution of each country has been to the final value/price of the product, which is largely determined by supply-demand considerations in the country of final destination. It will be very interesting to see, in the years after 2026, how the CBAM will be implemented and what will be the global reaction to border carbon taxes.

4.6.2 Is the Carbon Tax Regressive?

While the overwhelming majority of economists agree that a carbon tax may be necessary to speed up the transition to renewables, there is a worldwide debate on whether or not a carbon tax disproportionately distresses the poorer segments of the population – that is, a carbon tax is a regressive tax. A carbon tax is almost always passed to the consumers as higher prices for gasoline, heating oil, electricity, etc. In addition, since the manufacturing and distribution of almost all the consumer products entail energy expenditures, under the current energy consumption structure, the prices of most products will increase. The result will be a drop in the disposable incomes and the purchasing power of all households. From the macroeconomics point of view, an expected reduction of the citizens disposable incomes has detrimental effects on national economies, because lesser disposable income and purchasing power are associated with recessions, economic contractions, and higher unemployment – three undesirable societal effects [14]. In the absence of a rebate mechanism that will redistribute the carbon tax to the society, the tax will disproportionately affect the poorer segment of a country's population, which has less resources to pay for the higher prices of their electricity and consumer products.

Sweden was among the first countries to implement a carbon tax in 1991 [11]. The considerable revenue generated reverts to the general budget of the country and is not earmarked for specific projects. However, the general budget funds are used for specific purposes, such as addressing undesirable distributional consequences of the taxation (as lump sum payments to the lower-income households) and financing other climate-related measures (typically renewable energy projects) [15]. An early study by the Swedish Ministry of Finance, in 2003, determined that the carbon tax is regressive when measured against the annual disposable income of the Swedes [16]. A similar study, published by the London School of Economics in 2020, confirmed that the Swedish carbon tax is regressive when measured against the annual income of its citizens, but progressive when measured against their lifetime income. The last study also concluded that a carbon tax would be always regressive in countries with higher income inequalities (USA, Italy and France are included among the latter) [16].

A detailed economic study on the imposition of a carbon tax in the USA calculated its effects on incomes, assuming that all the tax revenue is recycled to the US citizens in one of three ways: a) as lump-sum payments to the lower two quintiles (poorer 40%) of the population; b) as a capital tax relief; and c) as a labor tax relief (e.g., lesser tax on wages). Table 4.2 shows the effects on the citizens that fall in five income quintiles with the 1st quintile (20%) being at the lowest income and the 5th quintile the highest income [17].

A glance at the data of Table 4.2 proves that only if the carbon tax is fully recycled with direct payments to the lower income citizens, the expected increase of the energy prices will have a positive effect on the lower three income quintiles. A capital tax rebate favors only the wealthiest classes, and a labor rebate does not favor any. The same conclusion was also reached in an earlier study for the USA, which determined that anything other than lump-sum rebates to the lower-income citizens would be regressive [18].

4.6.3 The Vicious Circle of Energy Taxes and Subsidies

Whether or not governments soften the impacts of a carbon (or any energy-related) tax by extending lump-sum payments and redistributing the tax revenue to the

TABLE 4.2
Effect of a Carbon Tax with Three Kinds of Tax Recycling in the USA as a Percentage of Income

Quintile	1st	2nd	3rd	4th	5th
Lump-Sum Rebate	3.36	1.29	0.3	–0.46	–1.93
Capital Tax Rebate	–0.87	–0.73	–0.57	–0.43	0.14
Labor Rebate	–0.28	–0.15	–0.18	–0.21	–0.45

Note: Negative values indicate income reduction.

lower-income households, an apprehension looms for the fiscal stability of the country when the energy transition is in its final stage, and the consumption of fossil fuels diminishes. If the renewable energy sources (in particular wind and solar with the associated energy storage infrastructure) fail to deliver energy at significantly lower prices, and the revenue from the carbon tax diminishes because fossil fuel consumption slows down, how will the nation continue to maintain the assistance to lower-income households and the current subsidies for the production of renewable energy?

When the energy transition is completed, the carbon tax revenue will dry. If the investment cost for renewables does not substantially decrease and energy prices do not fall, new revenue will have to be found to continue the energy subsidies. Another system of taxation will have to be imposed in order to continue the existing subsidies for wind and solar projects, and for payments to less affluent households. While such a taxation scheme may ensure that income inequality among households does not intensify as a result of the high cost of renewable energy, it will also reduce the disposable income of the middle class, with the consequence that more households will qualify for assistance and higher taxes may need to be imposed, thus fueling a vicious circle in the taxation-subsidies progressions.

4.7 DISTORTION OF THE ELECTRICITY MARKETS

The world needs reliable and affordable electric power when power is needed, not necessarily power from renewable sources that may fail to meet the demand. As the demand for electricity increases, reliable energy sources – such as fossil fuels, hydroelectric, nuclear, with the limited contributions by geothermal and biomass – are the only sources that supply dispatchable power and are capable to respond and satisfy the demand at any time. Carbon taxes and ever-increasing subsidies for wind and solar energy have distorted the energy investment field. Investment and production subsidies, in their various forms, can offset more than 50% of a renewable power unit's cost. At the same time carbon taxes (and the threat to increase taxation in the future) have discouraged investments in fossil fuel power plants within the developed countries. Such legislative and regulatory actions have tilted the electric power investment field in favor of wind and solar units, which are more profitable to build and operate than fossil-fueled power plants. The tilted playing field enables wind and solar corporations to underprice coal- and gas-generated electricity in competitive power markets. The often-observed negative electricity prices in the USA and European markets are a reminder that subsidies have distorted the electricity markets, which are not in harmony with the rest of the economy.

The uneven treatment of electricity generation sources has also had a "supply chain" effect on the manufacturers of gas turbines and coal power plants. Since utilities are highly encouraged (often mandated) to secure power from solar and wind generating corporations, gas turbine manufacturers have pulled back on investments for new gas turbines and (in 2026) there is an apparent shortage of new gas turbines, so that new gas power plants can't be built in time to meet the rising global power demand, especially the demand for new datacenters. This scarcity has encouraged arguments to offer more subsidies for wind and solar units. However, the gas turbine scarcity is entirely artificial and is due to the uneven energy investment field. Gas turbine manufacturers expanded production in the period 1998–2007 and accelerated turbine deliveries as the demand for new and efficient turbines increased. The same manufacturers were left with excess capacity when the gas turbine demand weakened since the 2010s, and the demand for wind turbines and photovoltaics skyrocketed as a consequence of wind and solar energy subsidies. The turbine manufacturers will, again, increase their manufacturing capacity for gas and steam turbines if they perceive that the investment playing field is level and their industry will not be hampered with excessive carbon taxes, governmental mandates, and subsidies for their competitors.

In market-oriented economies – such as the economies of most OECD countries – competition on a level-playing field is the cornerstone of the economic system. In the case of electricity generation, the playing field is tilted in favor of renewables and strongly favors wind and solar energy. As a result of governmental mandates, carbon taxes, and perpetual subsidies, the demand-supply equilibrium has been disturbed, and the electricity markets have been distorted to the point where consumers are not purchasing electricity generated by the lowest-cost primary energy sources, but they pay higher prices to conform to governmental mandates with the promise of GCC mitigation.

4.8 GLOBAL PRICES OF RENEWABLE ELECTRICITY

While predictions of future prices and costs of energy products are laden with substantial uncertainty, current and past prices are real and not uncertain at all. Current electricity prices hit the consumers' wallets every time they pay for their electricity to keep their lights on and their homes cool. One may look at the electricity prices paid by households around the world and find out what has been the effect on electricity prices in countries higher up in the transition ladder. When examining the cost of electricity from different renewable sources all the data show that electricity from hydroelectric power plants is not only more reliable, but also significantly cheaper than electricity from other sources including wind and solar energy. However, rivers with hydroelectric potential are not available in all regions of the planet, and most of the powerful hydroelectric resources have already been harnessed. Since the early 2000s, it is wind and solar energy that grew at astonishingly high rates. Their effect on current electricity prices offers a glimpse of the transition to renewables on the price of electricity in the future.

Figure 4.3 depicts the electricity prices paid by households of several countries in 2025 – in US$ per kWh – versus the sum of wind- and solar-generated electricity, as a percentage of the total electricity generation in the country [19]. The prices include taxes and distribution costs. It is apparent in Figure 4.3 that households in countries with high percentage of their electricity generated by wind and solar units pay substantially higher prices for their electricity. If one is to consider the trendline in the Figure, on average the electricity price increases by 0.72 cents for every 1% higher wind- and solar-generated electricity. The electricity prices are lower in countries with significantly high hydroelectric power production (Canada, Norway, Greece, P.R. China, Russia, which are denoted with solid square symbols). When

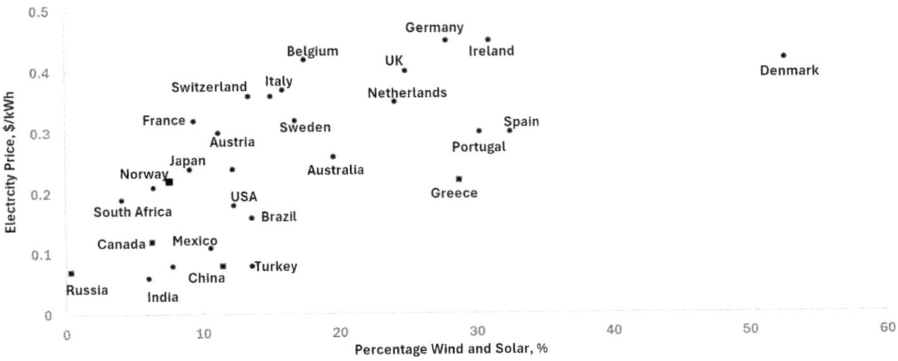

FIGURE 4.3 Household electricity prices in several nations in 2025 as a function of the wind- and solar-generated electricity. Square symbols (■) denote countries with substantial hydroelectric energy generation.

these countries are excluded, the trendline shows that the electricity price jumps to 0.87 cents for every 1% higher wind- and solar-generated electricity.

It is also apparent in Figure 4.3 that, in almost all countries, wind and solar energy contributed significantly less than 40% of the total electric energy consumed in 2025, so there was no need for substantial energy storage infrastructure to avoid power curtailment and the power surplus that leads to the duck curve or the rattlesnake effect (Figures 3.3 and 3.4). Denmark is the only outlier, with 53% of its electricity generated, primarily by wind. The excess electricity generated by the multiple wind farms in Denmark is exported to Sweden and Germany – a much bigger electricity market that can absorb large quantities of imported renewable energy. Because of the electricity exports to larger markets, only modest amounts of storage capacity are needed in the smaller and well-interconnected electricity markets, such as Denmark. The larger electricity markets (e.g., USA, Germany, P.R. China, India, etc.) do not have this advantage, since their excess electricity generation on windy and sunny hours can be too high to be exported to smaller markets.

4.8.1 Prices in the States of USA

In the recent years the GCC debate in the USA has become a "liberal vs. conservative" issue and since the country has essentially a two-party system, the GCC debate has become an election issue between the Democratic and Republican Parties. A slice of the liberal ideology, which has found its way on the platform of the Democratic Party, is that urgent action and immediate policies are needed to combat GCC. Within the Republican Party there are several views – from the extreme that the GCC is a "hoax," to more moderates that accept the reality of GCC but do not advocate immediate action, to those who consider urgent action. Matters such as energy transition, GCC, renewable energy, and cost of energy have become issues in State and Federal elections. As a result of this polarization, states that predominantly have Democratic Party governments (the so-called "blue states") have adopted and implemented policies favoring renewable energy, while states with Republican governments (the so-called "red states") have adopted milder versions of such policies and have given priority to generating inexpensive electricity for their citizens.

A group of eleven eastern states formed the Regional Greenhouse Gas Initiative (RGGI) in 2005 with the goal to reduce their CO_2 emissions by establishing a regional carbon cap, which is annually declining, thus imposing a limit on the emissions of power plants within the member states. Essentially the RGGI states (all of which have governments predominantly linked to the Democratic Party) adopted a cap-and-trade program, among other mechanisms, for the electricity industry within their states. The state of Pennsylvania (one of the original eleven states) exited the initiative in 2025, following the electorate's clamor to stop the rise of electricity prices. The state of Virginia originally entered the initiative, left it when a Republican governor was elected in 2022, and joined again in 2026 when a Democrat governor was elected – two events that underscore the political polarization of the GCC in the USA. Other blue states with Democratic Party governments adopted similar mechanisms to reduce GHG emissions via regulatory actions.

Figure 4.4 shows the average prices of electricity –including residential, commercial, industrial, and transportation customers – paid in 2025 in each one of the red states, the blue states, and the RGGI States (which are considered among the blue states) in cents/kWh versus the fraction of electricity generated by wind and solar [20]. Figure 4.4 reveals a few interesting trends for the electricity markets in the USA:

1. The blue states, on average, have substantially higher electricity prices than the red states. The average price for the RGGI states is \$0.196/kWh, in the other blue states, \$0.149/kWh, and in the red states, \$0.109/kWh. Alaska is the only red state with an electricity price higher than \$0.20/kWh, while the blue islands of Hawaii average a whopping \$0.38/kWh (energy, in general,

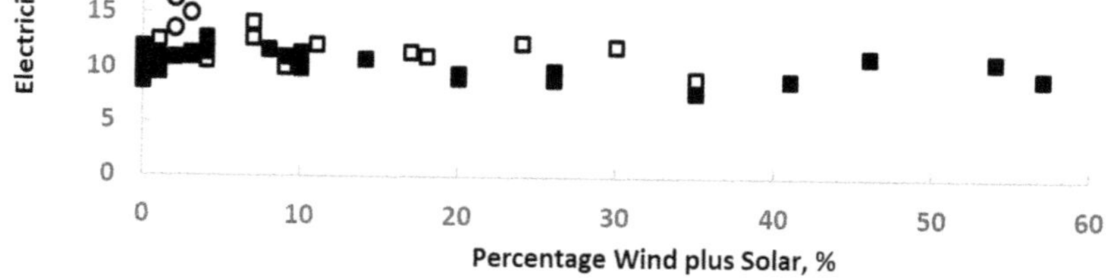

FIGURE 4.4 Electricity prices in the States of USA in 2025. The governments in the RGGI and blue States are considered more liberal, while those in red States are considered more conservative.

is more expensive in the remote States of Alaska and Hawaii!). The main reasons for the difference in electricity prices are:

a. State regulations and mandates that impose limits on fossil fuel generating capacity. The regulations compel electricity generation corporations in the respective states to add more wind and solar capacity regardless of the expected price to the consumers. An example is the case of Dominion Energy Corporation in Virginia. Following the *Virginia Clean Economy Act of 2020* to be carbon-free within a quarter-century and the specific mandate to build a massive offshore wind farm, this electricity-generating corporation contracted for the building of offshore wind capacity, fully knowing that this will increase the electricity price for its consumers [21].
b. High state subsidies for wind- and solar-generated electricity, despite the fact that solar irradiance and wind velocity may be limited (as in the north-east and mid-Atlantic regions of the USA).
c. High state and utility tax credits for the installation of roof-top PV panels, whose costs are ultimately absorbed by all the consumers [22].

2. The states with highest fraction of their electricity generated by wind and solar are primarily among conservative red states (Kansas, Iowa, Texas, and South Dakota) with favorable weather for wind energy generation. The electricity prices in these states are among the lowest.
3. The blue states with low electricity prices (and comparable to those of the red states) (Washington, Oregon, Illinois, New Mexico, Arizona) generate a high fraction of their electricity from hydroelectric, nuclear, or coal power plants.
4. None of the states depicted in Figure 4.4 has a significant energy storage capacity to support their renewable generation units. Similar to the case of Denmark, the smaller states (Kansas, Iowa, and South Dakota) that generate a great deal of electricity from wind do not need to develop energy storage capacity because they are connected to larger electricity markets and export any excess electricity.

It must be noted that wind- and solar-generated electricity enjoy subsidies in all the States by both federal and state agencies. Also, most of the states are interconnected with the several electricity grids that serve the entire country. Part of the electricity generated in one state may be sold in another at the prevailing prices and applicable taxes.

The data in Figures 4.3 and 4.4 support the conclusion that governmental control of the investments in the electricity generation sector – either by subsidies or mandates that favor certain energy sources and hinder others – may come to hurt the consumers as higher electricity prices. The claim that renewable tax credits, subsidies, and carbon taxation reduce the cost of electricity for households is contradicted by our current experience. While it is correct that wind- and solar-generated electricity are not burdened by fuel costs, the construction and maintenance of their generation units are still costly. In the absence of significant technological improvements that would substantially decrease the construction cost of wind and solar power plants, all indications are that electricity prices will keep rising with the higher penetration of wind and solar energy in the electricity markets.

Higher electricity prices (and energy prices, in general) are becoming an issue in the USA and European elections with the electorate demanding (and candidates promising) lower prices. This may mean that current subsidies will not be extended *ad infinitum* in the so-called "free-market economies" of the developed countries. A free market is the cornerstone of all modern democratic political systems. Energy markets, when allowed to operate freely, will provide adequate quantities of electricity, high reliability, and affordable prices.

4.9 NECESSARY INFRASTRUCTURE FOR THE ELECTRICITY GENERATION TRANSITION

When it comes to combating CO_2 emissions and GCC, all roads lead to the net-zero carbon transition for the global electricity generation industry – the substitution of all the coal and gas power plants to primarily wind and solar units. Nuclear energy is also an alternative, but existing policies and public sentiment (especially in the EU and the USA) diminish its contribution to the transition. The IEA has set this substitution as the first priority for the transition to renewables and has set a date, the year 2050, for its completion [23, 1]. In this section we will critically examine what is the required infrastructure for this transition, which at a minimum comprises the development of the following systems:

a. The necessary wind and solar (and perhaps nuclear) generating capacity.
b. The land required for the new generating units.
c. The necessary energy storage capacity to ensure reliability.
d. The overhaul of the distribution system (the physical electricity grid).
e. The maintenance of a few fossil-fueled units for increased reliability and emergencies.

Another scheme for net-zero, carbon capture and sequestration (CCS), is not considered as a viable option to be used in large-scale projects because it is too expensive; its pipelines require extensive licensing (especially when they cross state borders); and is already opposed by several environmental groups. Details on the disadvantages and high cost of CCS are included in Section 3.7.

To illustrate the magnitude and cost of the necessary infrastructure, calculations were performed with reference to the ERCOT electricity grid, which supplies most of the State of Texas, and for which daily and hourly operational data are available. The 2025 dispatchable generating capacity (capacity that is available anytime to satisfy the demand) of ERCOT included 4,975 MW of nuclear power, 15,500 MW of coal, and 68,800 MW of various types of natural gas units. The non-dispatchable renewable capacity of the grid included 40,570 MW of wind power, 35,120 MW of utility-level solar, and 11,000 MW of short-duration storage (mainly batteries) [24]. In 2025, the grid generated 488 TWh of electricity – more than the electric energy generated in the UK and Ireland combined, and more than that of Italy [25]. If the region supplied by ERCOT were an independent country, the region would have been the 6th largest electricity-consuming country in the world. This implies that the infrastructure for the decarbonization of ERCOT is comparable to the needed infrastructure of entire large countries.

4.9.1 Solar PV and Wind Generation Units

It is axiomatic that the substitution of fossil-fueled power plants with renewables must start with the construction of enough renewable energy units that are capable of supplying the needed power. Since geothermal energy resources are poor in most regions of the earth and are unable to economically supply a significant fraction of the electricity demand, and since most of the promising hydroelectric energy projects have already been developed, it will have to be wind and solar energy that will make up for the bulk of this substitution.

Replacing the fossil-fueled electricity generation units with wind and solar units does not imply a MW-to-MW substitution. As explained in Section 3.3.5, the power operating factors (POFs) of wind turbines are substantially low, averaging 33% in Texas [26] and close to 20% in Germany [27]. Solar PV units have even lower POFs and average 21% in Texas (one of the premier regions for the development of solar electricity generation units on earth) and close to 10% in Germany. The implication of the low POFs for the two abundant RES types of electricity-generating units is that significantly more RES power capacity needs to be constructed for the substitution of the existing fossil-fueled and nuclear units.

Because wind and solar energy are not always available, energy storage (in high-elevation lakes, batteries, hydrogen, etc.) must also be developed to meet the demand during periods of calm winds, clouds, and nights. All energy storage systems entail charging and discharging stages, with both stages dissipating power. The substitution of fossil fuels with RES-generated electricity necessitates the generation of higher amounts of electric energy than the actual demand, to meet the dissipated energy in the storage-regeneration systems. A detailed study with reference to the development of a grid-independent community in Texas concluded that, when accounting for the lower POF of the renewable units and the energy storage dissipation, the substitution of 1 MW of fossil generating capacity requires the development of as much as 5.4 MW of solar capacity or 3.4 MW of wind (or a combination of the two) [28].

The region served by the ERCOT grid does not have significant hydroelectric and geothermal resources. In the absence of further development of large nuclear power plants (which is very likely, given the NIMBY sentiment in the region) the grid must rely on wind and solar units for decarbonization. Given the hourly power demand in the year 2025 in the grid and the currently existing wind and solar electric power capacity, detailed calculations were performed to satisfy the hourly demand of the consumers with a combination of additional wind turbines and solar PV systems. Figure 4.5 shows the results of these calculations for the additional wind and solar power capacity, in GW (1 GW is 1,000 MW). Every point in the graph represents a combination of additional wind and solar power required for the substitution of the fossil-fueled power plants in the ERCOT region. One can easily observe in Figure 4.5 that the needed renewable energy infrastructure is substantial, with hundreds of additional GW of wind and solar units needed to replace the fossil fuel power plants.

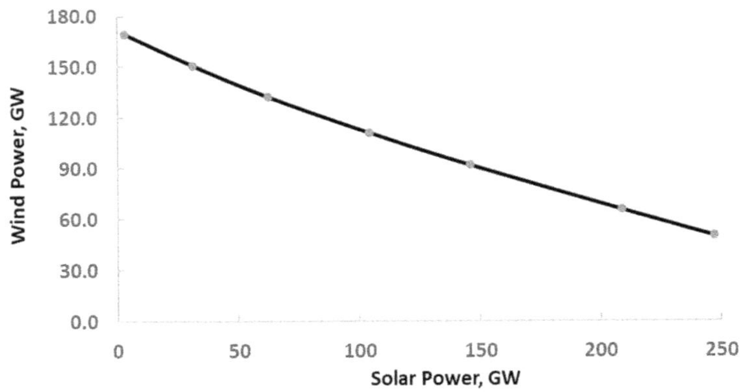

FIGURE 4.5 Additional renewable power capacity required for the decarbonization of the ERCOT grid.

4.9.2 Land for Wind and Solar Units

Nuclear and fossil-fueled power plants have substantially high power density – defined as power generated, in watts per unit land area used, in m^2. A nuclear power plant of 1,100 MW typically occupies a land area of 200,000 m^2 (50 acres) including its buffer zone, for a resulting power density of 5,500 W/m^2 [29]. The power density of coal power plants is similar, and that of the gas-fired units can be slightly higher. In comparison to nuclear and fossil fuels, the naturally available solar and wind energy resources are extremely diffuse (sometimes called dilute) energy sources, and the available power density is low. The *Blue Wind* PV solar farm in San Antonio, Texas – a prime location for high solar irradiance – features a power density of 30.3 W/m^2, while the *Gemasolar* thermal solar power plant in Spain has a power density of less than 10 W/m^2 [30]. Wind farms have even less power densities, at approximately 5 W/m^2, because the wind turbines must be sparsely spaced – at distances 6–10 rotor diameters from each other – so as not to interfere with the wakes of other turbines [31]. Moreover, the size of wind farms is limited to 20–60 MW in a single location to avoid significant interference of the turbine wakes.

The lower power density of wind and solar resources implies that large tracts of land must become available for the construction of the required power capacities of the renewable energy units that will replace the fossil fuels. Coming back to the case of ERCOT, land computations were performed for the required land area that needs to be occupied if the renewable installation combinations of Figure 4.5 were to materialize. Figure 4.6 depicts the results of the computations for the needed land area, in square kilometers, again as a function of the installed solar power in the ERCOT region (wind supplies the remaining energy, following the wind-solar combinations of Figure 4.5). It is observed in Figure 4.6 that the total land area of the wind and solar units would be in the order of 30,000 square kilometers, and this is an area equal to the size of the entire country of Belgium (or of the State of Maryland in the USA). Not only this is a tremendously high area, but also the cost of purchasing or leasing it would be substantial. Since the cost of undeveloped land in Texas in remote

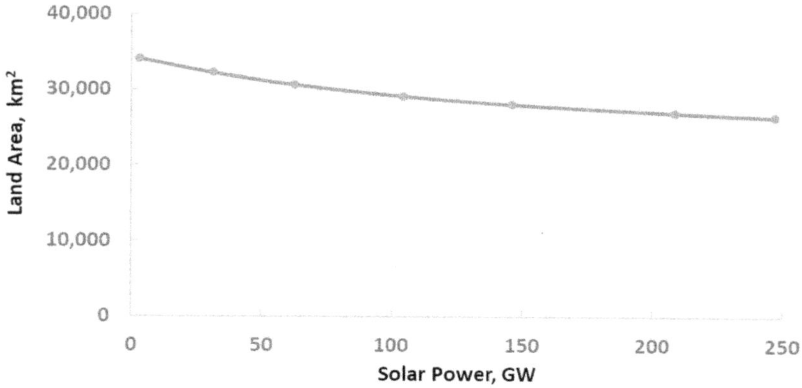

FIGURE 4.6 Land area required for the wind and solar installations that will substitute the fossil-fueled electricity generation in ERCOT. The area is of the size of Belgium.

areas is in the range of \$1,400-\$15,000 per acre (4,047 m^2) the cost of the land alone would be in the range 10–111 billion US\$.

4.9.3 Energy Storage

Energy storage is absolutely necessary for a reliable electricity grid powered by wind and solar energy. In the absence of adequate energy storage, households, businesses, and industries will experience blackouts during calm evenings and nights, with all the undesirable effects blackouts have. Section 3.3.4 lists the energy storage types of systems that are currently available or are in the research and development stages and may become available in the next two decades [32]. Geographic regions with hydroelectric potential may develop pumped-hydro storage (PHS) systems, which accounted for 94% of the global energy storage capacity in 2025 [33]. Almost all the existing hydroelectric power plants may also be used for energy storage after modifications to their reservoirs and their turbomachinery.

Regions that do not incorporate rivers in mountainous terrains cannot develop the large PHS systems and must rely on hydrogen, for long-term storage, and batteries, for short-term storage. The State of Texas, which is served by ERCOT, is one of the latter regions. Battery storage systems (including flow batteries) are simpler to construct, maintain, and use, but they self-discharge over long periods of time and are not suitable for seasonal energy storage (e.g., to store energy during the windy autumn and spring seasons to use it during the hot summers for A/C). Hydrogen, a stable gas, is ideal for seasonal energy storage, but the operation of hydrogen storage systems is significantly more complex. Hydrogen systems at a minimum comprise the following equipment:

a. Electrolyzers to generate *green hydrogen* from water and renewable electric energy.
b. Dedicated hydrogen compressors to compress the gas to a pressure in the range of 100–700 atm. This is because hydrogen is the lightest chemical element, and its density is extremely low at ambient conditions.

c. Robust hydrogen tanks to store the hydrogen at high pressure.
d. Fuel cells to convert the chemical energy of hydrogen back to electric energy.
e. Inverters and transformers to convert the direct current generated by the fuel cells to alternating current and then increase the AC voltage for transmission. Battery storage also needs inverters and transformers.

Calculations for the ERCOT grid to decarbonize and generate electricity solely from RESs show that both the required energy storage capacity and the energy dissipation due to storage are significant. Figure 4.7 depicts the two quantities, again as a function of the solar power capacity in the grid (wind supplies the remaining energy, following the wind-solar combinations of Figure 4.5). The total annual dissipation may be higher than the required capacity (the "size" of the storage system), because every hour of the year the storage system either charges or discharges at the expense of some energy.

It is observed in Figure 4.7 that the required storage capacity is substantial. If all energy storage is in hydrogen systems, the approximately 45,000 GWh of required storage entails 45 million m^3 of hydrogen tanks at 500 atmospheres pressure, dispersed in the region of ERCOT. If solely batteries are used, then 45 million large utility-scale batteries (of 1 MWh, each) are needed. An interesting fact here is that, if the latter batteries are of the lithium or lithium-ion types, the mass of lithium in the required batteries would be close to 540,000 tons – more than twice the total global production of this metal in 2024, which was 240,000 tons [34]. Another observation in Figure 4.7 is that the annual energy dissipation is also significant. At its minimum point the dissipation is equal to 8.6% of the annual electricity consumption in the grid.

It must be noted that the energy storage system of the grid should be designed to respond quickly to the demand fluctuations and the wind/solar supply fluctuations. The system should fast store all the supplied excess energy (supply minus demand) to avoid *renewable energy curtailment* (loss of energy generated by the renewable

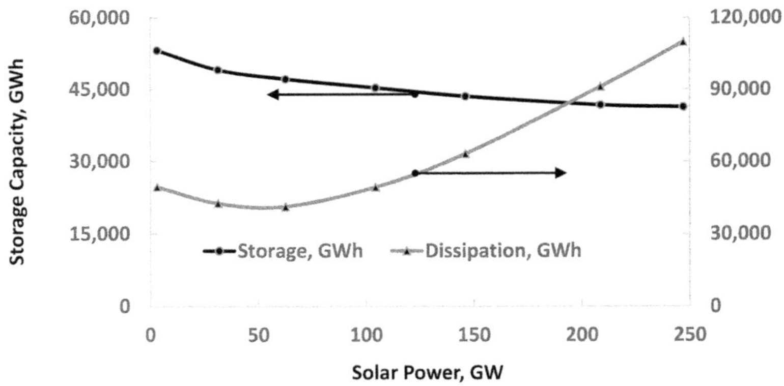

FIGURE 4.7 Required energy storage capacity for the ERCOT grid to decarbonize in the left axis. The annual energy dissipation is depicted in the right axis.

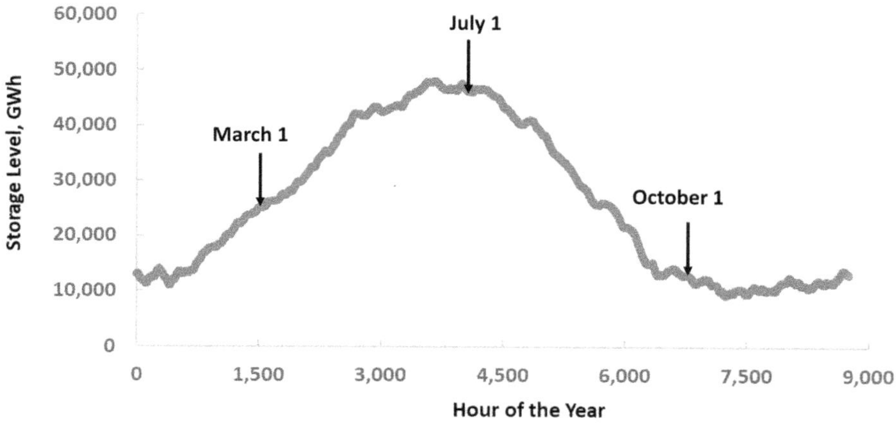

FIGURE 4.8 The level of energy storage for an entire year, starting at the first hour of January and ending at midnight on December 31.

units) and to also fast discharge energy to the grid to avoid large frequency and voltage fluctuations. The storage system should also be able to store energy for long periods of time. In the case of the ERCOT grid, where electricity demand is high in the summer, the energy storage system should be able to store energy during the late winter and spring, when the sun is bright and the winds stronger, in order to supply the region during summer and early fall, when the energy-voracious A/C systems kick in and the demand for electricity surges. This storage pattern is shown in Figure 4.8, which depicts the level of storage during all hours of the year, starting at the first hour of January and ending at midnight on December 31. For reliability, and to respond to weather extremes (including hurricanes, tropical storms, and polar vortices) at all hours of the year, the system holds sufficient energy to satisfy the grid demand for the following 168 hours (one week).

When it comes to energy capacity and energy storage, it must be noted that the internet is full of stories (verifiable and unverifiable) of heroes who managed to invent or develop significant energy storage with little investment. The author of this book always receives numerous such emails from his students. For example, in January 2026, a story was circulated of "...a man powered his home for eight years using 1,000 old laptop batteries..." together with a picture of a regular single-family house. When the story was researched it appeared that: a) the house pictured was not the one that was furnished with the battery system; and b) given that new laptop batteries hold 0.040–0.090 kWh, the system of one thousand batteries (40–90 kWh total capacity) in the claim would have been insufficient to power for a single night a mid-size household in the ERCOT region, which needs more than 100 kWh on hot summer nights. To top this, if homeowners were to purchase these batteries as new (there is not a market of millions of refurbished laptop batteries anywhere on the planet to be sold), the cost of the thousand batteries would be in the range of $100,000 to $250,000, an astronomical cost for any homeowner. One has to be very careful about stories and claims in the world wide web!

4.9.4 Transmission and Distribution Overhaul

Long-range electricity transmission networks are designed to minimize the dissipated power loss by transmitting power at high voltage – in the range 115–765 kV in the USA, 110–800 kV in Europe, and up to 1,000 kV in China. The transmission voltage is reduced by a series of "step-down" transformers in the substations and further reduced before the power is delivered to the households – at 120 V in USA, 100 V in Japan, and 220 V in Europe and China. A complex and intricately designed system comprising a multitude of generators, voltage inverters, transformers, and transmission lines, facilitates the transmission of power – at constant frequency and almost constant voltage – to households, commercial businesses, and industries.

Most of the currently operating fossil-fueled and nuclear power plants are located close to the major population centers, where the higher electricity demand is. However, good wind and solar energy resources do not follow the people, as fossil fuels have done for two centuries. Premium wind and solar resources are not necessarily located close to large population centers, so new transmission networks must be developed to bring the RES-generated electricity to the population centers. The costs of the new transmission networks are significant and, of course, they are passed on to the consumers as higher-priced electricity.

The island of Britain is an example of this happening. Following governmental mandates, all the coal power plants were shut by 2024, and wind energy was promoted (it is well-known that Britain is not located in a prime solar irradiance zone, so wind is the better alternative). However, the best wind resources are in the north, in Scotland, with meager transmission capacity to the major population centers, which are located in the south. This creates a bottleneck in the power generation-consumption scheme, which can only be remedied with a new and large transmission network along the north-south axis of the island. Since new transmission lines take long times to construct (and since in addition to all the other delays, local communities strongly object to high-voltage lines over their properties), it takes 5–10 years for a new wind farm in Scotland to join the electricity grid and sell power to the south of Britain. Apart from the transmission line costs, the connectivity delay also increases the costs for the generating corporations. With the passing of the generation and transmission costs to the consumers, electricity prices have almost tripled in the period 2010–2025 to the chagrin of the public [35]. For comparison, during the same period the cumulative inflation in Britain was 65%.

The story of Britain will be repeated in all parts of the globe, where the RESs are far from the population centers. In the case of ERCOT – where the premium wind resources are in the center-north of Texas (the Panhandle of Texas) and the premium solar resources are in the south and west parts of the State – tens of thousands of miles of new, high-voltage transmission lines will have to be developed for the ERCOT grid to become decarbonized at significant expense for the grid operators and higher prices for the consumers.

In addition to new high voltage lines and connections, RES-generating grids must counteract problems arising from lesser *grid inertia*, the result of lesser kinetic energy contributed by the massive and fast-spinning turbogenerators of fossil-fueled power plants. Loss of spinning reserves and inertia diminishes the grid's capacity for rapid frequency response and was the main cause of the blackout in Spain, on April

28, 2025 (see Section 3.3.3 for details). PV cells have zero spinning inertia, and the wind turbines rotate too slowly to provide significant spinning reserves. To counteract the loss of spinning reserves from traditional generators with renewables, the grid must develop fast-discharging battery storage systems; maintain a large number of *synchronous condensers,* rotating stabilizer equipment, such as older steel turbines and flywheels; add grid-forming inverters; and incorporate advanced control systems as well as demand management. These technologies offer support to quickly compensate for frequency and voltage fluctuations that follow the transition to high wind and solar power. Of course, they add to the cost of the decarbonized grid.

4.9.5 Emergency Fossil-Fuel Units

A well-managed electricity grid will always keep a few fossil-fueled power plants operational for enhanced reliability. Wind speed and solar irradiance highly depend on the weather, and weather extremes are expected to be more frequent with the GCC. During severe weather events (e.g., hurricanes), wind and solar generating units – which are located, by design, outside containment buildings and are exposed to the weather events – may get damaged and become unable to generate any electric power. To avoid regional blackouts, the emergency fossil-fueled units (in particular, the gas-turbines that do not contribute as much to CO_2 emissions as the coal power plants) will operate for a few weeks, until the health of the grid is restored. The emergency units will also help when the long-term demand grows at a faster rate (e.g., because of data center development or population growth) than newer RES units can be brought in line.

4.10 THE COST OF DECARBONIZATION

The last section and Figures 4.5 through 4.8 clearly show that the decarbonization of the electricity industry is not a simple task to be undertaken hastily. The substitution of the fossil fuel power plants that currently generate dispatchable electricity requires enormous infrastructure in new wind and solar units, large tracts of land, massive energy storage systems, and substantial alterations and additions to the electricity distribution networks. In free-market economies the development of this type of infrastructure requires substantial investment by the electricity generation and distribution corporations. The investment is recouped by higher future electricity prices. Even if the government undertakes to underwrite such decarbonization projects by increasing the national debt, the debt and the interest on the debt will need to be repaid by higher taxes on the citizens (the electricity consumers).

Let us assume that investment capital can be made available and that the stakeholders of the ERCOT grid decide to decarbonize the grid. The "good news" for such an endeavor is that the technology to transform the grid to net-zero carbon emissions is available, and the suppliers of the various types of equipment are in place. The "bad news" is that the process of timely decarbonization will encounter several bottlenecks, chief among which is the availability of materials and parts for the required energy storage systems (e.g., lithium for batteries, strong hydrogen tanks, electrolyzers, membranes for fuel cells, etc.). Let us further assume that such obstacles can be overcome and the manufacturers can produce enough PV cells, wind turbines, inverters, transformers, utility-scale batteries, hydrogen tanks, transmission lines, etc., to develop the new net-zero electricity generation and distribution network. The question that arises is what the cost of this transition will be, and how much electricity should be priced for the corporations to recoup their investments.

Based on the 2025 prices for the PV panels, wind turbines, land area, and storage systems [36], the cost estimates for the transition to renewables are in the range \$1.25 to \$2.00 trillion for the solar-wind power generating capacity combination that entails minimum storage in Figure 4.7 (the cost increases with higher storage capacity). The cost excludes any subsidies, since all governmental (State and Federal) subsidies must eventually be paid by someone – typically by current and future taxpayers. Following the electricity market practices, the decarbonization cost is recouped by the corporations that would participate in the decarbonization investment as the stream of annual revenues generated from the sale of electricity to the consumers. The corporations discount all future revenues by a discount factor – typically between 8% and 15%, a variable that is related to the internal rate of return and depends on the investment type and the investing corporation. For a life cycle of thirty years for the equipment used in the transition of the grid to RESs, Figure 4.9 depicts the electricity cost to the consumers (in US\$ of 2025 per kWh) as a function of the discount rate. The low estimate curve corresponds to the lower value of the required investment (total \$1.25 trillion), and the high estimate curve corresponds to the high estimate of the investment (\$2.00 trillion).

A glance at the electricity prices of Figure 4.9 proves that for the ERCOT grid to achieve complete decarbonization, the future electricity prices must increase substantially to levels that are similar to or above than those observed in the northern European countries in Figure 4.3. Such prices are several times higher than the electricity prices paid by the households in 2025 (\$0.142/kWh on average). For example, with a 10% corporate discount rate (a rather low rate for several utility companies), the prices that will need to be paid by the consumers are in the range \$0.47-\$0.68/kWh, or 3.3 to 4.8 times higher than the 2025 average household electricity price in Texas. Getting rid of fossil fuels and CO_2 emissions is an admirable and noble goal,

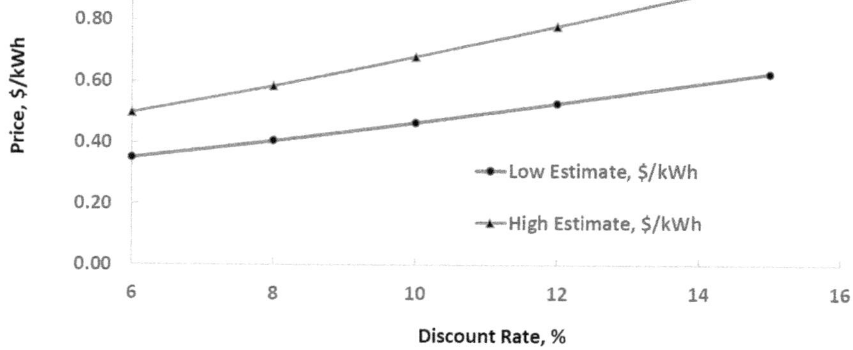

FIGURE 4.9 Electricity price (in 2025 US\$) that will enable electricity corporations to recoup their investment for the decarbonization of the ERCOT grid.

but is also expensive and implies that consumers will pay significantly higher prices for their electricity in the future.

It is noted that the calculations which led to the results depicted in Figures 4.5 through 4.8 and the prices in Figure 4.9 were conducted using the 2025 hourly electricity demand in the ERCOT grid (for 8760 hours total). One may argue that this demand may significantly drop using energy efficiency measures and retrofitting the buildings in Texas. If this is the case, less investment is needed by electricity-generating corporations. However, replacing equipment with more efficient ones and retrofitting buildings requires substantial investment by their owners. A 2020 study in Cambridge determined that retrofitting homes in the UK with the goal to reduce by 80% their CO_2 emissions, would cost approximately £135,000 per home (173,000 US$ of 2020) [37]. To do the same in the ERCOT region in 2025, the required investment would be in excess of 2.5 trillion US$, and still the utilities will have to invest and eliminate the remaining 20% of emissions. Decarbonizing a region is an expensive proposition that will be paid one way or another with higher investment or higher prices by the consumers and homeowners.

4.10.1 Carbon Dioxide Stabilization and Global GDP

Another way of looking at the cost of avoiding GCC is to examine what will be the cost of stabilizing the atmospheric CO_2 concentration. If getting rid of CO_2 emissions outright is very expensive, one may seek to find ways to stabilize the concentration of this gas in the atmosphere, while still emitting some CO_2 to avoid significant global economic disruption. A pertinent computational study was conducted for the cost of atmospheric CO_2 stabilization using eight economic models and reporting the average estimate of the models [38]. Figure 4.10 shows the results of this study for

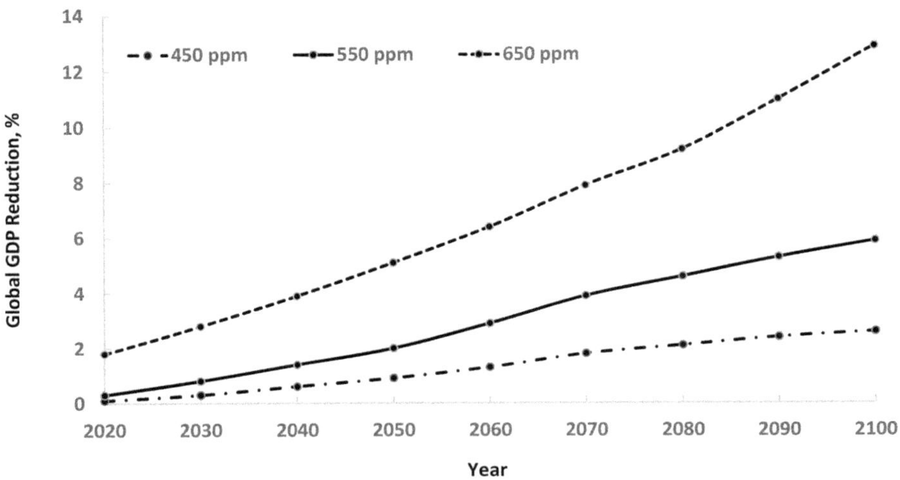

FIGURE 4.10 Cost of atmospheric CO_2 stabilization at the targets of 450 ppm, 550 ppm and 650 ppm, as a percentage of the global GDP.

the stabilization of atmospheric CO_2 at 450 ppm, 550 ppm, and 650 ppm. One may observe in Figure 4.10 that stabilizing the atmospheric CO_2 concentration at the level of 650 ppm may only cost 2.6% of the global GDP until the beginning of the 22nd century – about one year's GDP growth. Stabilization at 550 ppm, approximately twice the historical concentration of 280 ppm, will raise this cost to almost 6% of the global GDP, while stabilization at 450 ppm (note that it was at 429 ppm in January 2026) will cost close to 13% of the global GDP until the beginning of the 22nd century. The author of this study also notes that the most ambitious target of 450 ppm is unfeasible according to four of the eight models that were tested [38].

4.11 ENERGY PRICES, POVERTY, AND INEQUALITY

While the decarbonization of the economy – starting with the electricity generation sector – is desirable to mitigate GCC, this goal comes at a high price, both at the national and the global scale. The high prices will disproportionately affect the poorer households within a nation and the less affluent, developing nations.

4.11.1 Poverty and Inequality Within a Nation

High electricity prices have the same effect on individuals and households as high carbon taxes. Electricity is a necessary good for households, almost as essential as food is. Paying more for electricity implies that households will have less funds to pay for other goods, in particular non-essential goods. Significantly higher electricity prices affect substantially more the poorer part of the population, who will have less disposable income, than the more affluent classes. Like carbon taxes that are highly regressive without a redistribution scheme, high electricity prices hit hardest the poorer fraction of the households, unless governments take remedial action to provide subsidies for the poorer segments of the population [16, 17, 18]. However, unlike the carbon taxes that supply governments with revenue to redistribute, the higher electricity prices do not yield any revenue to be recycled to the poor. The higher energy prices are the result of the investment funds for the transition to net-zero carbon. When zero-carbon becomes reality, the investment has been spent, and the electricity-generating corporations (which typically assume significant debt for their projects) will collapse if the future prices do not increase to the levels that will enable them to recoup their investment and pay their debt.

Governments can still subsidize the high electricity cost for the poorer households but, since they do not have a parallel source of revenue, governments must tax another group (the wealthier electricity consumers) to make up for the subsidies. This implies that the wealthier consumers will bear doubly the burden of the net-zero transition and, since they are typically more frequent voters, such a taxation scheme becomes a politically undesirable option. Financing the subsidies with national debt is also undesirable because the debt will have to be paid in the future with higher taxation.

There is another unwelcome consequence of significantly higher electricity prices for the national economies. Because all the consumers will have to spend more on their electricity, they will have less disposable income to buy other economic goods, and the sectors of the economy that produce non-essential and luxury goods will severely suffer. Elementary macroeconomics predicts that this will cause the contraction of national economies and will lead to recessions. Recessions disproportionately affect the poorer parts of the population that have lesser resources to endure the economic hardships.

It has often been advocated by decarbonization proponents that, perhaps, we do not need as much investment in energy storage if we manage the demand side by *smart meters*. Smart electricity meters are installed in the households, measure the electricity use in real-time, and offer the present electricity price/cost to the consumers. When demand is high and supply low, the price increases, and the consumers decide whether to purchase the needed electricity and pay the asking higher price or switch off appliances (especially the energy-voracious A/C systems) to avoid an expensive purchase. On the other hand, when there is high supply (as happens during sunny and windy days) the price on the smart meter is low and consumers may switch on their appliances – and perhaps run their dishwashers and washing machines, which have been waiting for cheap electricity. Since the early 2000s smart meters have been part of households worldwide.

Managing the household demand with fluctuating prices often results in skyrocketing prices, which cannot be afforded by the poorer households. The regressive effects of such a pricing system have been seen in California, which is not even close to net-zero carbon emissions, but where households had to pay more than $1,000 per month in 2024 [39]. Residential customers in southern California had to pay $0.73/kWh in the late summer afternoons. In the northern part of the State, the electricity rates during the late afternoons were $0.56/kWh. The rates for consumers in San Diego jumped to $1.16/kWh during the so-called "Reduce Your Use Event" days. These high time-of-use electricity prices significantly affected the households in the hotter, inland regions, where summer temperatures routinely exceed 100°F (38°C), and A/C is essential. During extreme electricity cost periods, it is the poorer households that will switch off lights and appliances and, most likely, their A/C systems, an effect that may be succinctly expressed [39] as "make the poor sweat in the dark!"

Clearly high electricity prices (and energy prices, in general) have a disproportionate and adverse effect on the poorer households, do not help reduce poverty at all, and significantly increase the income inequalities within a nation. All in all, in the absence of meaningful national plans to subsidize electricity for the poorer households – in this case the question is where the funds for the subsidies will come – the higher electricity prices will exacerbate the already deep chasm between affluent and poor, and, very likely, will increase the poverty rate within a nation.

4.11.2 Inequalities among Nations

Let us now analyze the paths to decarbonization for developing nations. With very few exceptions (notably P.R. China and India) developing nations do not have a nuclear industry for electricity generation, and their manufacturing sectors for the massive production of PV cells, wind turbines, batteries, and hydrogen tanks are either weak or non-existent. To decarbonize its electricity generation industry, a typical developing nation will have to import most (if not all) the equipment for energy generation systems and energy storage systems as well as make the necessary improvements on its electricity distribution network.

Sizable investment will have to be made available and spent on the construction of these systems. The investment may not be of the size needed for the decarbonization of the ERCOT region ($1.25 to $2.00 trillion) but it can run into tens of billions of US$ for the grids of smaller nations and several hundreds of billions of US$ for the grids of larger nations. A factor that inflates the size of the needed investment is that the governments of most developing nations have set as their goals higher electrification and higher industrialization. This is also reflected in the 7th SDG (ensure access to affordable, reliable, sustainable, and modern energy for all). Higher electrification and industrialization entail significant demand growth in the next 20–30 years that will need to be met by the decarbonized electricity industry. However, the capital markets in most developing nations do not have the capacity to raise the tens or hundreds of billions of US$ required for this type of investment. The funds must be granted or borrowed from the developed nations. Actually, most developing nations (including the highly populated nations of India, Nigeria, Indonesia, and Mexico) have made their nationally determined contributions in support of the Paris agreement conditional on the availability of capital, other financial resources, technology transfer from the developed nations, and assistance in the development of renewable capacity-building support [40].

For most developing nations the infrastructure development for a decarbonized electricity grid exceeds their national debt – which is already sizable. The funds that currently become available for the decarbonization of the developing world are less than 100 billion US$ annually and most are in the form of export credits and loans from the World Bank that need to be repaid [41]. Given that the USA has withdrawn from the Paris accord; that several industrialized countries experience their own financial problems; and that a portion of the currently contributed funds are designated for "loss and damage;" it is not certain that the funds will become available year after year, and it is highly unlikely that the contributions will reach the targeted level of 300 billion US$ by 2035, as stipulated in COP-30 [42]. Even if that target is reached, simple arithmetic proves that global decarbonization is not possible by 2050 unless all developing nations take substantial amounts of additional debt.

Several industrialized countries – e.g., P.R. China, the EU, Japan, Canada, Australia, Korea, and the USA – may raise internal debt to pay for the needed decarbonization infrastructure that is manufactured within their borders. For them decarbonization can become an economic stimulus, justifying the debt. The 400 billion US$ *Inflation Reduction Act* of 2022 in the USA [43] and the $100 billion *Clean Industrial Deal* of 2025 for Europe [44] are aimed at strengthening the clean technology industries of the respective countries and maintaining industrial competitiveness against global rivals, while helping to reduce industrial CO_2 emissions. For the developing countries, which must purchase most of the equipment abroad, additional foreign debt is an anathema. Foreign debt needs to be serviced with periodic payments, and this entails outflows of funds that would have otherwise stimulated their national economies. Defaulting or restructuring the debt invariably leads to severe recessions and often to economic collapse – as it recently happened in Lebanon, Sri Lanka, and Zambia. The developing world found themselves in a *debt trap* in the late 1960s, which brought severe recessions, the stagflation of the 1970s and early 1980s, and precipitated economic, political, and social crises [45]. Funding their decarbonization with massive foreign debt to pay for equipment purchased from the industrialized countries will throw the economies of developing nations in a downward spiral.

As in the developed world, expensive electricity in the developing nations will hit harder the poorer segments of the population, who may not afford to buy this form of energy in the quantities they need. While the northern European households have the ability to pay the \$0.35-\$0.45/kWh, the vast majority of the households in Vietnam, Mozambique, and Brazil do not have this ability. The poorer citizens in the developing world will have to drastically cut their electricity consumption and (like the poorer citizens with smart meters in the developed nations) will have to "sweat it in the dark."

In summary, the developing nations lack the capital and the technology to reach zero-carbon emissions, even to significantly curb their emissions. Using their scarce resources to purchase the equipment and technology needed for decarbonization will further impoverish them, and the gap between poorer and more affluent nations will widen. In addition, this type of resource employment for the developing nations will disproportionately affect their poorer classes and will widen income inequalities within their borders.

4.12 THE HIJACKING OF THE PARIS AGREEMENT – REFLECTIONS ON COP30

Despite its vague stipulations and lack of enforcement power, the 2015 Paris agreement (initially signed by 195 nations) has been the most promising global pact for the governments of the world to come together, adopt policies, and develop programs to significantly and meaningfully reduce the global CO_2 emissions. The subsequent COPs were supposed to monitor the results of the Paris agreement and report meaningful CO_2 emissions reductions from the signatory nations. However, instead of fostering intergovernmental collaborations and joint efforts to combat the GCC problem and reducing the future GHG emissions, several delegates (mostly from developing nations) perceive the GCC fight as a vehicle to leverage and obtain economic advantages for their countries. In the successor COPs national delegations have introduced concepts such as *environmental justice*, *restoration*, *social justice*, *loss, and damage*, which are at best peripheral (and at worst irrelevant) to timely combating the increasing global temperatures.

Since representation in COPs is by nation (for all the signatory nations), and since consensus must be met for any decision, the developing world has an overwhelming majority in these Conferences and a very strong voice, which is heard throughout the globe. As a result, the focus on decreasing the GHG emissions has faded in favor of international aid and transferring funds to the developing world. A large part of the annual \$100 billion contributions, originally promised to promote renewable energy in the developing nations and mitigate the GCC, has become the *Fund for Responding to Loss and Damage* – essentially to compensate for damages from environmental disasters in developing nations. The annual "contributions" of the developed countries to this fund are slated to increase to \$300 billion by 2035. While it is uncertain who and by how much will contribute to the Fund, it is evident that the Fund has essentially become a tool for the transfer of wealth from the developed to the developing countries, without necessarily assisting their industrialization and robust economic growth. If these contributions are used to promote "social justice" and "loss and damage" the Fund's impact on the GCC will be minimal, and, given the lack of any planning, its impact on the reduction of poverty will be questionable.

Regarding tangible results to combat GCC, with the change of focus in the UN conferences and the diverse economic priorities of the governments represented in the COPs, it is no wonder that, between 2015 and 2025, the global CO_2 emissions have increased by 11%; the coal combustion (the fossil fuel that contributes the most to the emissions) increased by 13%; while the global population – a very strong contributor to the GCC as shown in Figure 2.5 – also increased by 12%. The lack of focus of the COPs on climate change has contributed to the derailment of the IEA goal for no more than 1.5°C globally-averaged temperature rise by 2050, as it is acknowledged in the 2023 report by this agency, which complains that although nations offer pledges to curb their GHG emissions, their current policies and measures do not reflect this goal [1]. Effectively, many nations proudly offer climate-benign pledges in the COPs, but they do not institute the policies to achieve their pledges.

COP30, the tenth successor to the Paris Conference, offered a sad reminder of this change of focus and proved that a large fraction of the nations represented have other priorities than curbing their CO_2 emissions. Held in the gateway to the Amazon ecosystem – in the tropical city of Belém, Brazil – the Conference was attended by thousands of delegates, but significantly fewer national leaders than recent Conferences. Notably absent were the leaders of USA [46], P.R. China, India, and Japan (four countries among the five largest global economies), while the German Chancellor (the third largest economy) was strongly rebuked for making unfavorable comments about the venue (events proved his comments correct!). Even before the Conference, the U.S. Energy Secretary fired the first salvo by stating about COP30: "It's essentially a hoax. It's not an honest organization looking to better human lives." The statement was reinforced by the non-governmental organization *Transparency International,* which publicized that "...more than half of all delegation members at COP30 have withheld or obscured details of their affiliations, potentially concealing conflicts of interest and undermining trust in the COP process..." [47]. In effect, special interest groups, activists, paid attendees, and lobbyists are taking over this annual Conference, whose goal is to combat climate change, and steer away the proceedings of the Conference to their own goals.

Among several days of violent protests by outside groups (in this case by indigenous people of Amazonia), torrential rains, floodings that did not respect the Conference venue, and a fire on the last day that frightened the delegates and sent the proceedings into overtime (an extra day for the Conference), the delegates managed to come to a consensus and produced a *Global Mutirão,* (roughly translated as global cooperation) as the final document of the meeting [48]. It is a testimony of the change of focus of the recent COPs that the words *fossil fuels* do not appear at all in the final document/agreement among the participating nations; the gas *carbon dioxide* appears only twice; the word *renewable* appears only once; and the word *nuclear* is absent. On the contrary the word *fund* appears seven times; the words *loss and damage* appear four times; and the word *trade* appears five times – to reflect the strong opposition of several participating "delegates"to the European carbon border adjustment mechanism, which threatens high carbon-content exporters to the EU.

Such developments that interfere with the spirit and goal of the Paris agreement to reduce GHG emissions have prompted the BBC reporters to include in their final reporting [49] "…do COPs still have a point?" In short, it is impossible that the world will come to an agreement, achieve decarbonization, and limit the temperature rise to less than 1.5°C with the theatrics of the recent COPs. The structure and *modus operandi* of the recent COPs simply cannot deliver the magnitude of the serious changes that are essential to ensure the success of the Paris Agreement and reduce the GHG emissions within the agreed timescale.

An interesting, and rather disturbing development in COP30 is the adoption of the *Declaration on Information Integrity on Climate Change,* which calls for action to counter the impacts of misinformation, disinformation, denialism, and deliberate attacks on environmental journalists, defenders (defenders of what?), scientists, and researchers that undermine climate action and threaten societal stability. What is not clear in this Declaration is who will determine what disinformation and misinformation is, and who will guard us from the "villains and rogues" that disinform and misinform the public. The author of this book remembers the words of the Roman satirist Juvenal: *Quis custodiet ipsos custodes?* (Who will guard the guards themselves?). Is this declaration a call for censorship? Will any opposing views to some of the self-proclaimed "climate scientists" and "defenders" be silenced, and will books like this one be burned for failing to fall in line with the dogmas promulgated by COP delegates? GCC is a global problem that requires a global solution. It does not belong to a few governments and activists, whose goals have obviously shifted from combating the principal causes of the GCC – the CO_2 emissions and global overpopulation.

NOTES

1 IEA, 2023, *Net Zero Roadmap: A Global Pathway to Keep the 1.5°C Goal in Reach 2023 update.* IEA publications.
2 Michaelides, E.E., A New Model for the Lifetime of Fossil Fuel Resources, *Natural Resources Research*, 26 (2017), 161–175.
3 See https://www.eia.gov/dnav/pet/pet_pri_gnd_a_epmr_pte_dpgal_a.htm.
4 See https://tradingeconomics.com/commodity/lithium.
5 Report of the World Commission on Environment and Development: Our Common Future. Available at: http://www.un-documents.net/our-common-future.pdf.
6 See https://sdgs.un.org/goals.
7 Global indicator framework for the Sustainable Development Goals and targets of the 2030 Agenda for Sustainable Development https://unstats.un.org/sdgs/indicators/indicators-list/
8 See https://unstats.un.org/sdgs/report/2025/
9 US-DOE, Energy Information Administration, 2023, Federal Financial Interventions and Subsidies in Energy in Fiscal Years 2016–2022, IEA Report August 2023.
10 Coady, D., Parry, I., Le, N. P., & Shang, B., 2019, *Global fossil fuel subsidies remain large: An update based on country-level estimates.* International Monetary Fund, Report WP/89/19.
11 See https://www.government.se/government-policy/taxes-and-tariffs/swedens-carbon-tax/

12 European Commission, 2021, Proposal for a regulation of the European Parliament and of the council establishing a carbon border adjustment mechanism, pdf.
13 Lim, B., Hong, K., Yoon, J., Chang, J.-I., Cheong, I., 2025, Pitfalls of the EU's Carbon Border Adjustment Mechanism. *Energies*, 2021, 14, 7303. https://doi.org/10.3390/en14217303.
14 Samuelson, P.A. Nordhaus, W.D., 2009, *Economics,* (19th Edition), McGraw-Hill Education, New York.
15 See https://www.government.se/government-policy/taxes-and-tariffs/swedens-carbon-tax/
16 Andersson, J,, Atkinson, G., 2020, *The distributional effects of a carbon tax: The role of income inequality.* Centre for Climate Change Economics and Policy Working Paper 378/Grantham Research Institute on Climate Change and the Environment Working Paper 349. London: London School of Economics and Political Science.
17 Goulder, L. H., Hafstead, M. A.C., Kim, G. R., Long, X., 2019, Impacts of a carbon tax across US household income groups: What are the equity-efficiency trade-offs? *J. Public Economics*, 2019, **175**, 44–64.
18 Williams III, R. C., Burtraw, D., Morgenstern, R. D., 2015, The Impacts of a US Carbon Tax across Income Groups and States, *Resources,* **56**, 25–29.
19 Sources: https://worldpopulationreview.com/country-rankings/cost-of-electricity-by-country, https://ec.europa.eu/eurostat/statistics-explained/index.php?title=Electricity_price_statistics, and https://www.iea.org/reports/electricity-2024/prices
20 Data from https://www.eia.gov/electricity/sales_revenue_price/pdf/table_4.pdf
21 J. Glock, 2026, Virginians Will Pay, and Keep Paying, for "Affordable" Energy, *Wall Street Journal*, January 30.
22 Hiller, J. and Rust, M., 2025, Costly power bills spark outrage, *Wall Street Journal*, December 30.
23 IEA, 2021, Net Zero Roadmap: Net Zero by 2050 A Roadmap for the Global Energy Sector, IEA publications, France.
24 ERCOT-fact sheet, October 2025. Available at: https://www.ercot.com/files/docs/2022/02/08/ERCOT_Fact_Sheet.pdf.
25 See https://www.ercot.com/gridinfo/generation
26 See https://www.ercot.com/files/docs/2025/01/18/ERCOT-Monthly-Operational-Overview-December-2025.pdf
27 Smil, V., 2020, *Energiewende, 20 Years Later*, Spectrum, IEEE publ. December.
28 Michaelides E.E., 2025, Energy Efficiency and the Transition to Renewables – Building the Communities of the Future, *Energies*, 18(7), 1778.
29 Correia, J., 2012, Wind power to nuclear power infographic comparison, *Nuclear Newswire ANS*, February 9.
30 See https://solarpaces.nrel.gov/project/gemasolar-thermosolar-plant-solar-tres
31 Michaelides, E. E., *Energy, the Environment and Sustainability,* CRC Press, Boca Raton, 2018.
32 Michaelides, E.E., 2025, *Energy Storage – Driving the Renewable Energy Transition,* CRC Press, Boca Raton, Florida.
33 United States Department of Energy DOE Global ES Database. Available at: https://www.sandia.gov/ess-ssl/global-energy-storage-database-home/
34 See https://www.statista.com/statistics/606684/world-production-of-lithium.
35 Feng, R., 2025, Britain's Grid Can't Handle Green Push, *Wall Street Journal,* December 31.
36 Zakeri, B. and Syri, S., 2015, Electrical ES systems: A comparative life cycle cost analysis, *Renew Sustainable Energy Review* **42**, 569–596.
37 Kelly M., 2020, *Decarbonising Housing: The Net Zero Fantasy*, Global Warming Policy Foundation, Report #38.
38 Tol RSJ, 2010, Carbon Dioxide Mitigation, in *Smart Solutions to Climate Change – Comparing Costs and Benefits,* Lomborg B. ed., Cambridge University Press, Cambridge.
39 Lesser, J, 2025, California's Plan: Make the Poor Sweat in the Dark, *Wall Street Journal*, January 31.

40 Jordan, CM, 2024, Paris Agreement: Nationally Determined Contributions Conditional on International Support—In Brief, https://www.congress.gov/crs-product/R48248.
41 World Bank, 2025, *World Bank Climate Finance 2025,* World Bank, New York.
42 COP30, 2025, Global Mutirão: Uniting humanity in a global mobilization against climate change, CMA.7, November 22.
43 See https://www.motive-power.com/breaking-down-clean-energy-funding-in-the-inflation-reduction-act/.
44 See https://commission.europa.eu/topics/competitiveness/clean-industrial-deal_en.
45 Payer C, 1974, *The Debt Trap: The IMF and the Third World*, Penguin Books.
46 Under the Trump administration, the USA has exited the Paris Agreement.
47 Half of Cop30 delegates obscured details of affiliations, warns Transparency International, *The Guardian*, November 13, 2025.
48 COP30, 2025, Global Mutirão: Uniting humanity in a global mobilization against climate change, CMA.7, November 22.
49 BBC, November 22, 2025, also article on November by J. Rowlatt, November 9, 2025.

5 The Sustainability Conundrum

In light of the enormous price for decarbonization infrastructure and the future cost of the highly touted energy transition, this Chapter examines the consequences of a net-zero carbon electricity generation industry *vis-à-vis* the set of the four sustainable development goals (SDGs), which were identified in Section 4.4.

1. End poverty in all its forms everywhere (SDG 1).
2. Ensure access to affordable, reliable, sustainable, and modern energy for all (SDG 7).
3. Reduce inequality within and among countries (SDG 10).
4. Take urgent action to combat climate change and its impacts by regulating emissions and promoting developments in renewable energy (SDG 13).

With the firm conclusion that energy transition is incongruent with actions to eradicate poverty and reduce the inequalities within and among nations, the only realistic path for the future is the global adaptation to the effects of climate change and the prioritization of the SDGs, without abandoning the CO_2 emissions mitigation efforts. Providing the world's population with sufficient and affordable energy is the key to the SDG prioritization strategy. The chapter ends with observations on twelve trends on energy demand and supply that will endure in the era of climate adaptation.

5.1 THE SUSTAINABILITY CONUNDRUM

Given the high decarbonization costs to the consumers, the 7th SDG, to ensure affordable and reliable sustainable energy, contains an internal contradiction: because viable and economically feasible geothermal and hydroelectric resources are limited – actually, non-existent in most parts of the globe – the vast majority of the countries will have to rely on the non-dispatchable wind and solar energy to gain access to sustainable energy. The reliable generation and distribution of electric power derived from wind and solar energy requires the acquisition of vast tracts of land and the development of costly infrastructure – new wind and solar units, energy storage, grid enhancements, etc. Such endeavors necessitate large investments, which result in significantly higher future electricity prices. As it is apparent in Figure 4.3, the current data on the global electricity prices (which do not yet include the cost of the massive energy storage systems) show that countries and regions with electricity grids that rely on high fractions of wind and solar energy for their electricity, already charge significantly higher prices.

The 13th SDG is partly correct in that to combat the "climate crisis," the world needs to promote higher use of renewable energy. However, a major driver of the

 DOI: 10.1201/9781003674658-5

climate crisis is overpopulation. There is no mention of this compelling issue either in the 17 sustainability goals or in the 169 targets to achieve the goals. The second part of the goal, to regulate emissions (presumably those of greenhouse gases, including CO_2) can backfire if it deprives the global population of the less expensive sources of energy and results in lack of energy when needed, or non-affordable energy. Our modern society – indeed, our civilization – was built on reasonably priced energy and heavily depends on the supply of adequate and affordable energy quantities for all. National and local regulations that simply restrict the use of energy or make energy prices unaffordable for the common people will have devastating effects on the social structure.

The 1st SDG to end poverty in all its forms everywhere is a noble goal, worth undertaking in all nations. However, this goal clashes with the 7th and 13th goals, which essentially imply the promotion and development of wind and solar energy at the exclusion of fossil fuels. All calculations and indicators stemming from current prices lead to the conclusion that electricity is more expensive when it is generated by the non-dispatchable wind and solar power plants and will be substantially more expensive when energy storage systems are needed to ensure reliability. In the absence of significant energy subsidies for the poor, high electricity prices will reduce the disposable incomes of all consumers and will hit particularly hard the poorer households. Currently (2026) energy subsidies for the poorer households in most countries are largely derived from carbon taxes imposed on fossil fuels. In a decarbonized future the carbon taxes vanish (or severely diminish) and the tax revenue is not there anymore. The continuation of energy subsidies for a large fraction of the population raises the question of who will pay for the subsidies when a society achieves its zero-carbon goal, and carbon taxes disappear.

The 10th SDG to reduce inequality within and among countries also clashes with the 7th and 13th goals for the development of an infrastructure to generate the more expensive wind and solar electricity. The preceding paragraph covers the reduction of poverty and inequality within a nation. All the data indicate that, if energy becomes more expensive, the gap between rich and poor will increase. Charging consumers different prices with smart meters to follow the availability of wind and solar power, deprives poorer households from using electricity when they really need it, and effectively "makes the poor sweat in the dark." Such practices instead of reducing income inequalities actually enhance inequalities within countries.

Let us now consider the second part of the 10th SDG, the reduction of inequality among countries. Most of the developing countries do not have the technology and the relevant industry to internally manufacture and set in place the necessary infrastructure for the decarbonization of their electricity sector. PV cells, wind turbines, advanced utility-scale batteries, and fuel cell membranes are manufactured by corporations in a handful of countries to be exported. Most developing nations need to import the decarbonization equipment at the expense of scarce capital that, otherwise, could have been used to enhance industrialization and improve their economies. Since capital is not readily available – and the grants for the decarbonization of the developing world are meagre – individual nations will have to resort to external debt to pay for the net-zero carbon infrastructure. Even if we assume that the loans will come with low interest from the World Bank (which is largely financed by the developed countries), first, the Bank does not have the tens of trillions of US$ required for the loans; and second the developing world is already awash in loans – several of them in default, restructured, or simply non-performing. To assume additional debt with the expectation of repayment will increase the dependence of the developing nations on the more affluent ones. In addition, the allocation of scarce capital for equipment importation will slow down the road to industrialization for the developing world. The result will be slower economic growth for the developing countries (or zero growth), thus increasing the current inequalities among nations.

It is not a coincidence that developing countries, which prospered at the beginning of the 21st century and shrank their distance from the affluent countries, did so with the development and consumption of domestic energy sources – typically fossil fuels that increased the global CO_2 emissions. The Peoples Republic of China essentially relied on its coal resources to develop its manufacturing sector. Between 1990 and 2019 P.R. China increased its coal production by 266% and managed to increase its GDP by 556%, while increasing its CO_2 emissions by 320% [1]. Saudi Arabia relied on its abundant petroleum reserves. In the same period Saudi Arabia increased its domestic crude oil consumption by 271%, increased its GDP by 488%, and its CO_2 emissions went up by 214%. India's recent economic development was also built on its coal reserves. Between 1990 and 2019 India increased its coal production by 355%, to enjoy a GDP growth of 613% (despite some problems with its volatile currency) and its CO_2 emissions went up by 312%. While these three countries – and several others that narrowed their GDP gaps with the affluent world – also developed some of their wind and solar resources, the bulk of their electricity generation, the bulk of their general energy consumption, and their recent prosperity are inextricably tied to their domestic energy resources, chiefly fossil fuels.

All in all, the high cost for decarbonization and the higher electricity prices will affect an entire nation (or region in the case of ERCOT) and will penalize more the poorer households and the less affluent nations. This means that the poverty levels will increase and the purchase power inequalities among the groups of the population will also increase. Debt-laden and less affluent nations will have to delay their industrialization and economic growth to import the decarbonization infrastructure, with the result that the GDP gap among rich and poorer nations will increase.

Effectively, the transition to net-zero carbon electricity generation, which is the consequence of the 7th and 13th sustainability goals, will disproportionately affect the less affluent nations and poorer citizens within all nations; it will increase the poverty levels worldwide; and will set back the path to achieve the 1st and 10th sustainability goals. This is the conundrum the SDGs present the humanity of the 21st century, an intricate and difficult problem with an apparently impossible solution.

5.2 EPILOGUE

The fact that the United Nations Framework Convention for Climate Change (UNFCCC) and the United Nations Conference on Sustainable Development (UNCSD) have adopted a set of goals that are internally inconsistent does not mean that the goals should be abandoned. Eliminating poverty, improving health, justice for all, and a clean environment are noble goals worth striving for. What is missing is the prioritization of the goals according to the corresponding threats to humanity and the urgency to combat the threats. Eliminating poverty and inequality, while striving for a near-term future (2050) with solar and wind energy at the exclusion of fossil fuels, are mutually incompatible goals. Instead, the world may continue its quest to improve the lives of the poor and level the playing economic field for all nations, while adapting to the effects of global climate change and taking steps to wean itself from fossil fuels in the far future. The following sections form a path to an energy-secure future and include actions we can take to mitigate the effects of the GCC and slowly wean ourselves from fossil fuels.

5.2.1 Adaptation

For many years the environmentalist community and climate activists have framed the debate around climate change as an existential emergency for humankind requiring immediate and radical action. Some went further and declared that the planet is "in a death spiral." Self-described "experts" and "celebrities" around the world clamored that we are in the middle of a planetary climate emergency, that battles for food are soon to come, and that the third world war is in the near future. Expressions like "security threat," "war," "battle," have speckled many speeches and essays to underline the need for urgent action. Even the 13th of the SDGs uses language with warlike connotation: "...to combat climate change."

For several activists and "celebrities" adaptation may have the meaning of defeat in the "war to save the planet." However, in this case, adaptation doesn't imply defeat or surrender but survival for a large part of the earth's population. We have reached a time in the history of the climate debate when exaggerations with combative overstatements should be abandoned and realism should be embraced. Common people around the world have shown that they favor the higher living standards (indeed affluence!) and the economic security that comes with affordable and reliable energy, rather than constant climate. With the knowledge that the vast majority of the governments – despite the ominous predictions – are unable or unwilling to take drastic action and urgently "combat" climate change, it is time for our world (and the UN agencies) to develop a long-term strategy and take meaningful actions to ensure the safety and future prosperity of the world's population.

Pausing the war for climate change and entering an era of a climate truce and cooperation does not mean that the efforts of the last thirty years were in vain. Policies were enacted, technology was developed, and the energy transition was partly implemented in several nations. In 2026 wind and solar energy generate a sizeable fraction of electricity – 9% in the USA, more than 25% in China, close to 15% globally. As a result, substantial CO_2 emissions avoidance has been achieved and the global CO_2 emissions per unit economic activity have consistently dropped for over 50 years, as shown in Figure 5.1, which depicts the amount of CO_2 emitted globally per thousand US$ of economic activity [2]. Such good news, however, are overshadowed by the robustly growing energy demand, which is driven by a stubbornly increasing global population that strives to achieve prosperity. Given the reluctance of most CO_2-emitting nations (especially the two larger ones) to subscribe to binding commitments, the global CO_2 emissions are likely to increase for one or

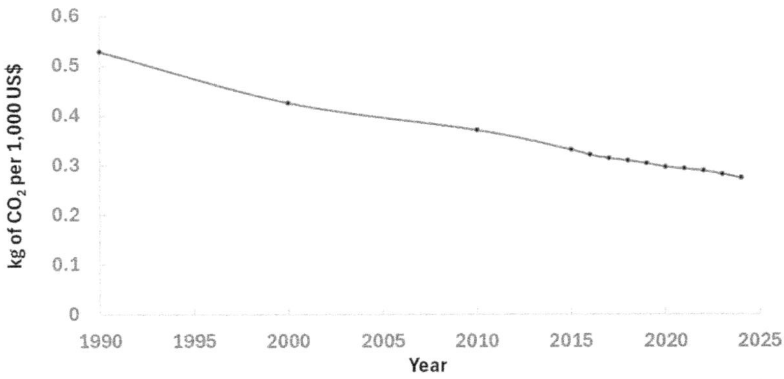

FIGURE 5.1 Global CO_2 emissions per thousand US$ of economic activity. Higher efficiency and more renewable energy reduced the emissions by 48%.

two decades and then stabilize for another decade, before they retreat. This implies that the atmospheric CO_2 concentration will reach higher levels during the middle of the 21st century and the average temperatures will rise more than 2°C. The higher global temperatures will bring excessive heat in the summer but milder winters, occasional floods in low-lying areas, and more summer illnesses transmitted by mosquito bites, but less pulmonary diseases inflicted by the severe winter cold.

Climate adaptation does not carry the exorbitant bill of net-zero carbon by 2050 and, to a large extent, is feasible for most. Air-conditioning (A/C) of homes and offices is ubiquitous in the developed countries and well in progress in most parts of the developing world. Other heat adaptation measures, such as better insulation in older buildings, high insulation in new buildings in combination with more efficient A/C systems, have already been written in the building codes and standards of most urban areas. When the summer heat adaptation measures (efficient A/C systems) are combined with winter heat reduction measures (insulation) their benefits may be higher than their costs for the homeowners.

National investment in flood protection measures (seawalls, levees, and dams) for vulnerable coastal regions, alongside planning by individual owners, such as land-use planning, and property-level improvements, will alleviate the effects of the (highly probable) sea-level rise and storm intensification. An early study by the World Bank [3] quantified the incremental costs of activities to make projects resilient to increased global temperatures as 5 to 20 percent of the portfolio at risk, or at least $1 billion per year of additional World Bank funding. The total cost of vulnerable infrastructure adaptation may amount to 10–100 billion US$, significantly smaller than the required infrastructure for decarbonization.

National health efforts against the vector-borne diseases (from ticks and mosquitoes) can relieve the suffering of the affected populations by eradicating the carriers and vaccinating the population [4] while reaping the benefits of less cold-afflicted illnesses [5] (mainly cold-related respiratory and cardiovascular illnesses), which currently kill more people in the developed world [6]. While it is always a difficult task to estimate future health care costs, the study by Ebi [4] offers a range of 4–13 billion

US$ for adaptation measures that would adequately respond to tropical diseases if the atmospheric CO_2 concentration rose to 750 ppm (compared to 429 ppm in 2026).

Agriculture is another sector that is likely to be affected by increasing global temperatures. Adaptation practices in this sector are decided by the farmers and are implemented at the microscale of the farm, or the village. Typical adaptation strategies by farmers include the crop mix and the timing of planting and harvest. Central governmental agencies can help with improved irrigation and water conservation systems at very low cost. An earlier report by the IEA estimated the annual expenditures in this area to be 7 billion US$, with more than 80% of this sum dedicated to new irrigation systems [7]. Two bright consequences for the agricultural sector are that the higher global temperatures will open a great deal of land in the northern hemisphere (in particular, in Canada and Russia) to agriculture, and that plants grow faster and become more robust at higher CO_2 concentrations. GCC and higher temperatures will not deprive the global population of sufficient quantities of food. The forecasted "battles" and "wars" for food by the "experts" are extremely unlikely to materialize in a future with adaptation.

Regarding national security in an increasingly volatile world, the USA National Intelligence Council warned [8] that "...climate change will increasingly exacerbate risks to US national security interests as the physical impacts increase and geopolitical tensions mount about how to respond." The Council classified the risks as geopolitical tensions, geopolitical flashpoints, and country-level instabilities. Such risks for the USA national security (and several other OECD nations) could imply the need for security agreements with other nations, projecting limited power in volatile regions, and preparing for human migration. While the author of this book is not an expert on national security, he is certain that defense departments and security services around the world with annual budgets of hundreds of billion US$ can find efficacious solutions to this risk for the developed nations. As for "energy security" governments around the world will have to decide and build their own energy mix, based on local and imported energy sources that include both renewables and fossil fuels.

5.2.2 Continuation of Mitigation Efforts

While the world embarks on an adaptation course, mitigation efforts should not be given up. Reducing the CO_2 emissions is essential to limit the global temperature rise and the scale of the GCC problem. Also, it should not be forgotten that the quantity of fossil fuels on the planet is finite and that there will be a point in the far future when further extraction of fossil fuels becomes more expensive than wind and solar energy. The investments in renewable energy, energy efficiency, and conservation of the past fifty years have resulted in creative solutions for many global problems. These solutions can help when eventually the carbon-based resources in the planet will be exhausted.

The 21st century saw the remarkable growth of renewable energy for the generation of electricity. While, before the year 2000, hydraulic energy was the dominant form of renewables for electricity generation, by 2024, the combined generation from wind and solar units has surpassed the hydroelectric generation. The 21st-century trends may be seen in Figure 5.2, which depicts the annually generated electric energy, in

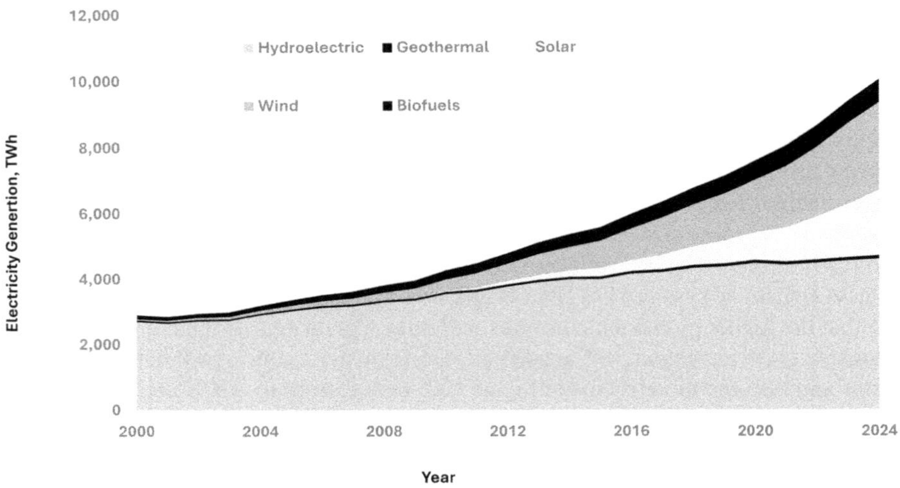

FIGURE 5.2 Electricity generation by renewable energy sources in the 21st century. The contributions of solar and wind power have grown remarkably.

TWh, from hydraulic, geothermal, solar (both PV and thermal), wind, and biofuels. The annual rate of growth of electricity generated by solar energy is a roaring 34.1%, and that of wind-generated electricity is 19.1%. Even though these stellar growth rates may decline in the future, it is worth generating as much electricity as the two sources may economically provide. One may also see in Figure 5.2 that hydraulic and geothermal energy had rather slow growth rates (under 3% annually) primarily because their economic harnessing is confined to specific parts of the globe.

5.2.3 Prioritization

A glance at the seventeen sustainable development goals (SDGs) proves that, while all the goals are noble and worth pursuing, the limitation of financial resources dictates that the global community cannot pursue all of them simultaneously. Prioritization becomes necessary. Securing *affordable and reliable energy* is the key to achieving all seventeen goals. *Sufficient and affordable energy* should be the first priority of developing nations. The development of the now affluent nations started in the Industrial Revolution, when energy sources (first wood and then coal) became the motive power of the industrial engines. Reliable and affordable energy ushered mechanized agriculture and liberated the population from working the land. Thermal engines, powered by the reliable fossil fuels as energy sources, manufactured the industrial goods that contributed to affluence and leisure. This gave rise to the development of the middle classes that are the backbones of democratic societies and the impetus for innovation and prosperity in today's developed economies.

Clean water should be a goal that needs to be prioritized. In 2026 one of eight people does not have access to clean, drinkable water. Diseases ravage populations that lack clean water. Without clean water agriculture does not produce foodstuff. However, water (in most places impure and non-potable) is not a scarce resource. Rivers, lakes, and ponds are abundant on earth; and the oceans embrace all the continents and make saline water available to the vast majority of nations. Readily available energy can provide clean water for agriculture and potable water for people and animals. Less than 2 kWh of electricity will convert one cubic meter (1,000 liters or 264 gallons) of ocean water to clean water for agriculture or drinking, and about 0.8 kWh will convert one cubic meter of brackish water to potable water. The availability of affordable energy and clean water is a boon to agriculture. Sufficient *food* can be produced to nourish the populations and eliminate hunger and food insecurity globally. Since *poverty* is always associated with food insecurity, poverty rates will also drop.

In the decades that followed the Industrial Revolution, thermal engines in combination with reliable and sufficient energy created long-sustained economic growth, *employment* opportunities, and *decent work* for individuals. This can be repeated in the developing nations of the 21st century to create new employment opportunities and decent work for their citizens. Since we have been taught the history lessons of the past, in the 21st century we can avoid the social mistakes of the rampant 19th-century industrialization and aim for an inclusive and humane global industrialization that follows the procurement of sufficient and cheap energy.

It is axiomatic that hungry people do not care a lot about their *environment.* Securing sufficient food is their first priority. The availability of sufficient food, which is produced by the affordable and plentiful energy and water, the elimination of hunger, and the creation of decent jobs will boost what we loosely call "the middle classes" in all nations and will also turn the attention of populations toward their environments. People will look after improving their living conditions and nurturing their environment. A well-fed population with sufficient energy will take better care of the human settlements (both cities and countryside) and of the global environment.

The lessons from the history after the Industrial Revolution also show that affordable and reliable energy, which helps liberate humans from tilling the earth, provides them with clean water and sufficient food, and creates decent employment opportunities, will also foster the other objectives and characteristics touted in the seventeen SDGs. More people will have the opportunity to be educated. More nurses, medical doctors, and other health professionals will see to it that diseases are eliminated and public health is improving. Education will also promote democracy within nations, will help achieve gender equality, and (why not?) will help stabilize the population of this planet to a sustainable level that will not threaten the terrestrial ecosystems and the marine environment.

All in all, reliable, affordable, and sufficient energy is the key to gradually achieving all seventeen sustainability goals. Affordable and reliable energy was the linchpin that created the now developed world and will be the essential element for the development of the other nations, so that we may *eliminate poverty* on a global scale and *reduce inequality within and among countries* in the future.

5.2.4 A Dozen Energy Trends in an Adaptation Era

Like many other economic sectors, the energy sector is laden with uncertainty, and this complicates the work of policymakers, businesses, and common citizens.

Geopolitical tensions and pressures in long-established partnerships and alliances catapulted the *world policy uncertainty* and the *world trade uncertainty index* [9] to stratospheric heights in September 2025, with little subsidence in 2026. However, in this time of high uncertainty, the author of this book discerns a dozen trends that will shape the global energy demand and supply.

1. Electricity will be the dominant "fuel" of the 21st century. Electric vehicles, manufacturing of specialized materials, sensors, data centers, and A/C in a warming climate are the drivers for significant electricity growth. Electricity generation is growing at more than twice the growth rate of the other end-user energy forms and will continue well into the middle of the 21st century.
2. Energy efficiency and conservation efforts are indispensable in a growing electricity future. The current substantial investment by both the governmental and the private sector, the promising research in academic institutions, the progress of artificial intelligence, and the further automation of industrial tasks will result in significant energy savings.
3. Wind and solar energy will continue growing at fast rates, despite the expense associated with their investment. The new wind and solar generating units have almost met the growth part of the electricity demand in 2025 and are expected to continue doing so with double-digit growth rates in the near future.
4. The growth of wind and solar electricity generation units will be followed by increased energy storage capacity. Grid stability and power arbitrage at the utility scale, as well as protection from blackouts in households, are the drivers for energy storage. Electrochemical energy storage (batteries) and pumped-hydro systems will dominate this growth. Hydrogen will take a back seat unless we have significant technological breakthroughs to substantially improve the efficiencies and reliability of fuel cells and electrolyzers.
5. Geothermal and hydraulic energy can only be economically harnessed in specific regions on the planet. Even though they generate dispatchable electricity and are preferable to wind and solar, their annual growth rates are not expected to significantly exceed the recent 2–3% rates.
6. Under the pressure of the environmental community, the use of coal for electricity generation will start declining in the 2030s. National policies all over the world are detracting investors from new coal power plants, and several of the older plants are scheduled to close. However, the industrial use of coal (particularly in the cement and steel industries) will very likely continue unabated.
7. A great deal of the void created by the closures of coal power plants will be filled by natural gas and the highly efficient *combined cycle gas turbines*. Natural gas generates dispatchable electricity, and the gas turbines may be switched on and off in short times (within 2 to 5 minutes) [10]. As such, they are ideal complements for the less consistent wind and solar power generating units.

8. Despite the clamor for their elimination, petroleum and liquid hydrocarbons will still be widely used for transportation. The pervasiveness of electric vehicles notwithstanding, gasoline and diesel engines will always be more convenient for long-haul transport, and indispensable for airplanes and ships (both of the passenger and the freight type).
9. Banning a disaster similar to the Chernobyl accident, nuclear energy will come back to prominence. Since 2015 nuclear energy has been growing at a faster than historic rates and in 2026 there are more than 70 GW of new nuclear capacity under construction. By 2050 the global nuclear capacity may expand to as much as 1,446 GW [11]. While the larger reactors (more than 1 GW) are expected to count for most of this growth, the development of small modular reactors (in the scale of 0.1 GW) could be the answer to the soaring electricity demand of the energy-voracious data centers.
10. States are becoming the dominant forces that shape the energy markets, rather than leaving this function to businesses and free markets. National and regional policies will continue to dictate (to a large extent) the energy mix for electricity generation and the growth of primary energy source consumption.
11. Energy security risks are expanding from the traditional petroleum and natural gas supplies to the stability of national electricity grids, power lines, and strategic minerals for photovoltaics, wind turbines, and electricity generators [12]. Rising security risks now include cyberattacks and malicious activity targeting critical infrastructures.
12. Emerging nations will drive the global energy movements. As the traditional energy players – primarily the OECD countries and P.R. China – enter an era of relative economic stability and energy equilibrium, a group of emerging economies – India, Indonesia, Vietnam, the Middle East, South America, and the developing African nations – will increasingly shape the global energy markets.

In summary, the world's population can take prudent measures to adapt to the increasing temperatures; can ensure the supply of reliable and affordable energy that will help with the further economic development of all nations; while at the same time, it can promote the reasonably priced renewable energy sources (and, perhaps nuclear) to prepare for a stable and more prosperous future.

NOTES

1 Our World in Data. See https://github.com/owid/co2-data?tab=readme-ov-file
2 Data from https://github.com/owid/co2-data?tab=readme-ov-file
3 World Bank (2006), *Clean Energy and Development Towards an Investment Framework,* World Bank, Washington DC.
4 Ebi, K.L. (2008), Adaptation Costs for Climate Change of Diarrheal Disease, Malnutrition and Malaria to 2030, *Global Health*, 4(9) doi: 10.1186/1744-8603-4-9.
5 Tol, R.S.J. (2009), The Economic Effects of Climate Change, *Journal of Economic Perspectives*, 23(2), 29–51. doi: 10.1257/jep.23.2.29

6 Amirkhani, M., Ghaemimood, S., von Schreeb, J., El-Khatib, Z., and Yaya, S., (2022), Extreme Weather Events and Death Based on Temperature and CO_2 Emission: A Global Retrospective Study in 77 low-, Middle- and High-Income Countries from 1999 to 2018, *Preventive Medicine Reports*, May 31; 28,101846.
7 IEA (2007), *World Energy Outlook 2007*, IEA, Paris.
8 National Intelligence Council (2021), *Climate Change and International Responses Increasing Challenges to US National Security Through 2040*, Report NIC-NIE-2021-10030-A, Washington DC.
9 See https://worlduncertaintyindex.com/
10 U.S. Energy Information Administration, Annual Electric Generator Inventory.
11 *2026 World Nuclear Energy Outlook*, World Nuclear Association, London, UK. See https://world-nuclear.org/images/articles/World-Nuclear-Outlook-Report_dfed5656.pdf.
12 EIA, *World Energy Outlook 2025*, EIA publications, Paris, France.

Index

Note: Page references in italics denote figures.

For Product Safety Concerns and Information please contact our EU representative GPSR@taylorandfrancis.com
Taylor & Francis Verlag GmbH, Kaufingerstraße 24, 80331 München, Germany

www.ingramcontent.com/pod-product-compliance
Ingram Content Group UK Ltd.
Pitfield, Milton Keynes, MK11 3LW, UK
UKHW020128300926
476127UK00009B/444
9781041144809